Pvt. Ltd.

An E-learning
Approach to Teach Discrete Mathematics

Pramila Ramani
Assistant Professor
Department of Education
Central University of Tamil Nadu – 610 005

P. Thiyagarajan
Assistant Professor & Head In-charge
Department of Computer Science
Central University of Tamil Nadu– 610 005

2019

Studium Press (India) Pvt. Ltd.

An E-learning Approach to Teach Discrete Mathematics

ISBN: 978-93-85046-47-6

Published by:

Studium Press (India) Pvt. Ltd.
4735/22, 2nd Floor, Prakash Deep Building
(Near Delhi Medical Association)
Ansari Road, Darya Ganj, New Delhi-110 002
Tel.: + 91-11-43240200-15 (15 lines); Fax: 91-11-43240215
E-mail: pubdir@studiumpress.in

Printed at India

About the Authors

Pramila Ramani PhD (Educ); *Educationist; b* September 9, 1970 Tamilnadu; *m* Ramani Ramanathan; *Educ* M S Univ of Baroda; specialisation in: Educational Technology, Educational Administration, Research Methodology, Population Studies, Mathematics Education, Philosophy, Sanskrit Literature, Abstract Algebra, Real Analysis and Discrete Mathematics, Mathematics; Asst Prof, Education Dept, Central Univ of Tamil Nadu; Member in the Editorial Board: European Journal of Special Education Research, European Journal of Education Studies; Member in the Editorial Board & Reviewer, World Journal of Educational Research 2016-18; Board of Editorial/ Advisory Team of GJMS International Publication for three years from 2017; Editorial Advisory Board of Journal of Global Education and Research; *publs* — Efficaciousness of CAI in Teaching of Mathematics (2016), A Study of Two Year B.Ed Programme (2016), Cogency of Continuous and Comprehensive Evaluation in Mathematics (2016), Dynamics of Programmed Learning Material in Teaching of Mathematics (2016); *Awards* NavrachanaUniv Awards-Vibrant Gujarat Initiative 2014, Navrachana University Awards-Service to Science Education Program 2014, Teacher of Developing Era Award 2017; *Address* Education Department, Central University of Tamilnadu, Neelakudi, Thiruvarur 610005, India. E-mail:pramila.maths @gmail.com Tel.: 09488526627.

P. Thiyagarajan did his Ph.D. in Computer Science and Engineering and Post Doctorate from Pondicherry Central University and Department of Atomic Energy, Government of India respectively. His area of research includes Educational Technology, Threshold Cryptography, Information Security, Steganography, Big Data Analytics, Security aspects in Smart Cities and Cryptography. Currently he is working in Central University of Tamil Nadu, Thiruvarur 610005, India Email: thiyagu@cutn.ac.in; Tel: 09488584015

Preface

E-learning is imparting education through computer and network enabled digital technologies such as Internet, intranet, TV, audio, video resources etc. The term e-learning was coined in the seminar conducted in Los Angeles USA in the year 1999. In simple words e-learning can be defined as use of Information and Communication Technology (ICT) to enhance and support learning. In the recent past, Internet has become one of the fundamental need for human to lead their life. Internet is merged with all phases of human life including learning. Learning also takes place in Internet and it is called by various names such as web based learning, online learning, virtual learning etc.

The characteristic of e-learning such as learner-centric learning, lifelong learning, flexible learning social learning and personalized learning makes is more popular among user, such that the most difficult subject can be learned easily by the user.

In computer science, one of the tough subject to be marked by students is Discrete Mathematics. In this books four topics from Discrete Mathematics such as Permutations, Combinations, Equivalence Relation and Recurrence relation were selected for study and e-learning materials are developed for the above topics and complete analysis have been made on the e-learning materials developed.

This book is divided into six major sections namely: (i) Introduction to e-learning in Mathematics (ii) Detailed literature survey in the field of computer assisted instruction, e-learning in mathematics and education (iii) Details of the study that are followed in developing e-learning materials in Discrete Mathematics (iv) Research Methodology (v) Results and discussions and (vi) Future directions.

Section I gives on over view about e-learning its advantages, how it is used in mathematics, benefits of having e-learning in Mathematics.

Section II detail about the literature in the field of Computer assisted instruction, e-learning other than mathematics, e-learning in mathematics, problems in learning discrete mathematics and barriers in implementing e-learning in education.

Section III discuss in brief about the rationale of the study, conceptual framework, delimitation of the study, research questions and objectives of the study.

Section IV elaborates on research methodology, tools used for data collection and analysis.

Section V describes about interpretation of ANOVA, reaction scale and the discovery from the analysis of the effectiveness of the developed e-learning materials.

Section VI talks about the conclusion, acknowledgment and future directions of the research.

Table of Contents

Section 3

Section 4

Section 5

Section 6

List of Tables

List of Figures

Development and Implementation of E-Learning in Discrete Mathematics

ABSTRACT

Today is the world of technology and it has versatile usage. Students use technology gadgets for various purposes. Investigators tried to use this technology in teaching learning purpose. Investigators through their experience understood that computer science students find it difficult to learn discrete mathematics. The use of e-learning is as follows. Students can watch e-modules and lectures; they can even download these lectures for reviewing. Such digitized content is very helpful to the learners and teachers as well. Students can learn in their own convenience (time) whenever they want to learn and wherever they want to learn (place). E-learning can cater all types of learners. It can also cater to different levels of learners like slow learners, mediocre learners and advanced learners. Experts can video their lectures and upload it in the suitable website. Different models can be shown through these video lectures thereby making learning experiences rich and effective. Therefore they developed e-learning material in discrete mathematics offered in department of mathematics and computer science and implemented the same in odd semester during the academic year 2017.The present study is delimited to universities in Tamil Nadu and only Discrete Mathematics offered in department of Computer Science and Mathematics.

Objectives of the Study

1. To develop e-learning in Discrete Mathematics.
2. To study the effectiveness of the developed e-learning in terms of students' achievement in Mathematics.
3. To study the reaction of the students about the mode of learning discrete mathematics at the end of the experimentation.

Population of the Study

All students studying in universities of Tamil Nadu in the department of mathematics and computer science constitute the population.

Design of the Study

The present study is an experimental study. The investigators followed pre-test posttest control design. Mathematics Achievement Scores was analyzed quantitatively using ANCOVA. Responses given by the students to the Reaction Scale of the experimental group were analyzed quantitatively using Chi Square and percentage. The analysis of ANCOVA showed that there is significance difference between mean marks of experimental group students and mean marks of control group students' at .05 level of significance. Findings from the analysis of reaction scale for experimental group indicate that out of **forty** statements the computed Chi-square values of **thirty three** statements were found to have statistically significant higher values than the tabulated value and the students were found to have **positive reaction** and **favorable attitude** towards the statements carrying such **higher values.** That is, for more than **50%** of the statements they responded **positively**. Using the result the investigators concluded that e-learning is one of the effective methods to learn discrete mathematics.

***Key words*:** e-learning material, Digitalized content, Teaching learning process, Different types of learners and Different modes of learning

---

Section 1

1. INTRODUCTION

Investigators developed e-learning material in discrete mathematics offered in department of mathematics and computer science and implemented the same in odd semester during the academic year 2017. They found the effectiveness of the developed material. (Wikipedia) **E-learning** (or e-learning) is the use of *electronic educational technology* in learning and teaching. Conceptually, e-learning is broadly synonymous with instructional technology, *information and communication technology* (ICT) in education, EdTech, learning technology, multimedia learning, technology-enhanced learning (TEL), computer-based instruction (CBI), computer managed instruction, computer-based training (CBT), computer-assisted instruction or computer-aided instruction (CAI), internet-based training (IBT), flexible learning, web-based training (WBT), online education, virtual education, *virtual learning environments* (VLE) (which are also called learning platforms), *m-learning,* and digital education. In usage, all of these terms appear in articles and reviews; the term "e-learning" is used frequently, but is variously and imprecisely defined and applied. These alternative terms are all linguistically more restrictive than "*educational technology*" in that they refer to the use of modern tools, such as computers, digital technology, *electronic media*, networked digital devices and associated software and courseware with learning scenarios, worksheets and interactive exercises that facilitate learning. However, these alternative names individually emphasize a particular digitization approach, component or delivery method. Accordingly, each conflates to the broad domain of educational technology.

2. E- LEARNING TO ENHANCE QUALITY IN EDUCATION

Many Indian universities are adopting these higher order techniques for example IITs and IISc (E-learning Courses From the IITs and IISc) use e-learning resources. Students can watch e-modules and lectures; they can even download these lectures for reviewing. Such digitized content is very helpful to the learners and teachers as well. Students can learn in their own convenience (time) whenever they want to learn and wherever they want to learn (place). This learning is independent of different cultures, different groups, and different socioeconomic backgrounds. It can even reach students from geographically distanced places where teachers cannot reach. Moreover, classrooms have heterogeneous groups of students from different backgrounds. (Alberta Education, Alberta, Canada, 2014) The learning environment should value and respect the diversity of students' experiences and ways of thinking, so that students are comfortable taking intellectual risks, asking questions and posing conjectures. Students need to explore problem-solving situations in order to develop personal strategies and become mathematically literate. They must realize that it is acceptable to solve problems in a variety of ways and that a variety of solutions may be acceptable. Therefore to teach them is a challenge. There are different types of learners like those who can learn through audio, visual, video, etc. E-learning can cater all these type of learners. It can also cater to different levels of learners like slow learners, mediocre learners and advanced learners. Experts can video their lectures and upload it in the suitable website. Different models can be shown through these video lectures thereby making learning experiences rich and effective.

3. IMPORTANCE OF MATHEMATICS

Epistemologically mathematics means mathema - explaining and understanding, tics- techniques such as counting, ordering, sorting, and measuring. Right from pre-historic period there have been problems to solve. Problems may be over basic requirements like food, water, shelter or accomplishment like constructing multi-storey building.

Now the Communications revolution that is sewing humanity, symbolized by internet and mobile cell-phones would not have been possible without a copious application of Mathematics. Sometimes, a branch of knowledge waits for the mathematical maturity, like Boolean Algebra for the development of digital circuitry and Computers. Once students who were mathematically weak used to opt for Economics till John Maynard Keynes and his cohorts raised the cream of Economists skimming with Mathematics. Now, to study Economics and management sciences one must possess stout Mathematical heart. Humans marvelled at the flight of birds or the fishes' swim but could only dream to replicate their movement. Only by using mathematics he realized his dreams. Now his aeroplanes and helicopters do a better job and submarines are more versatile than whales. All this is made possible through Mathematics.

In this modern era we cannot think of a field, where calculation or computation is not used. Knowingly or unknowing we use mathematics in our day to day life. It ranges from household to industries, business, education, science and technology, art and craft and even in music, dance etc.

Mathematics is part and parcel of daily life. Mathematics is used in learning almost all subjects. We cannot imagine learning engineering disciplines without mathematics. Biology, medicine, computer, science, economics etc. all use mathematics. The revolution in information and technology is due to advancement in mathematics. Statistics uses mathematics for analyzing of data. Different commissions have given different views on the place of mathematics.

Everybody uses mathematics whether they realize it or nor, for example a cook uses mathematics to modify amount according to the number of persons or to modify recipe. Shopkeeper uses mathematics to calculate change, tax, and sales prices. Travelers use mathematics to plan their trips, they calculate time of arrival and departure. Even homeowners use mathematics to determine the cost of materials when doing projects.

Professionals like accountants, agriculturists, architects, Biologists, Chemists, Engineers (Chemical, Civil, Electrical, Material, Industrial...) Geologists, Medical Doctors, Managers, Meteorologist, Military Personnel, Nurses, Politicians, Technicians, and Tradesmen- all use mathematics.

4. E-LEARNING IN MATHEMATICS

ICT (Information and Communication Technology) has great potential for teaching and learning process at all levels. The use of ICT has enriched the teaching learning process with the help of computer. It has brought a great change, innovativeness, and creativity in teachers in teaching learning process. Mathematics and computer are both important in today's life as they open the gate of ample opportunities in this modern world. Mathematics is widely used in computers both in hardware and software. Computer helps in improving the knowledge of mathematics. Computer helps in making classroom teaching lively.

Computer can play vital role in learning process as it can work with the imagination of students. Any concept in mathematics can be explained with the help of pictures and this visual image can help in understanding the concept at ease. In paper pencil method student can get bored easily and can find it difficult to practice the sum again and again. E-learning works as a change and increases the curiosity of students and they can learn interestingly without any difficulty. Also whatever is learnt through computer, the contents can be retained for longer time as they use more senses of the students. Varity of exercises can be provided and this ensures active involvement of the students. The material can be provided according to the needs of the students.

4.1. Advantages of E-learning in Mathematics

In this age of technology the teacher should be aware of student's need. E-learning can help to satisfy the needs of the students. E-learning lessens the workload of the teacher besides it has many other advantages as given below

Provides Wide Range of Experiences

E-learning helps teachers to provide a wide range of experiences/ he can give many examples and illustrations and can make the concept clear.

Provides Motivation

It can sustain the motivation of the students as the topic can be presented in an enjoyable manner as concepts can be presented systematically, interestingly and immediate feedback can be given which sustains the motivation of the students. Graphics and pictures can be presented which can attract and retain student attention. Students get reinforcement when they answer the question correctly and the topic is presented in a systematic manner in an increasing order of difficulty.

Individualized Instruction

E-Learning is an individualized instruction as it caters to the individual difference. Some students are slow learners and some are fast learners. Classroom is a heterogeneous group. Some students need more time to learn while others need less time, so learning speed differs from learner to learner. E-learning also provides different learning experience according to the understanding level of the students. It also provides facilities like selecting the topics of their own interest. It provides individual attention to each and every student and thus enhances the quality of teaching learning process.

Interactive Learning

E-learning provides immediate feedback to the students and thus constantly interacts with them. In E-learning students actively take part in the learning process. As it contains many examples and diagrams it makes the learning process interesting.

Section 2

5. REVIEW OF THE RELATED LITERATURE

Purpose of the Review

(Koul, 2008) stated the following purpose, "Review of the related literature; besides, allowing the researcher to acquaint himself with current knowledge in the field or area in which he is going to conduct his research, serves the following specific purpose:

- It helps the researcher to understand particular trend and locate the gaps in research areas of his/her interest.
- The review of related literature enables the researcher to define the limits of his field.
- By reviewing the related literature the researcher can avoid unfruitful and useless problem areas. "He can select those areas in which positive findings are very likely to result and his endeavour would be likely to add to the knowledge in a meaningful way".

Educational technology is an important aspect of educational system. It is extensively used to provide effective teaching learning process in the classroom. Realizing the importance of e-learning as an instructional tool various researches were conducted and its effectiveness was tested. In this section investigator has provided available related literature keeping the present study in view. A total of 8 studies have been reviewed to develop a holistic perspective of the objectives and findings of these studies and to arrive at the implications to support the present study.

1. The related literature in the field of Computer Assisted Instruction
2. The related literature in the field of e-learning other than Mathematics
3. The related literature in the field of e-learning in Mathematics

4. The related literature in the field of problems in learning discrete mathematics
5. The related literature in the field of e-learning/technology in discrete mathematics
6. The related literature in the field of barriers in implementing e-learning in education

5.1. The Related Literature in the Field of Computer Assisted Instruction

Antony (Rose Antony Stela, 1992) tested the Effectiveness of Computer Assisted Instruction with special reference to underachievers Ph.D. Education Bharathidasan University. Problem the study throws light on the application of CAI and the teacher support system (TSS) for the optimum development of underachievers (UA). Objectives (1) To develop CAI software (2) To find out the effectiveness of CAI with TSS and CAI without TSS with references to the learners variables *viz* sex, IQ and achievement level and (3) To find out the interaction of the learner variables and the treatment on the achievement score. Methodology: The randomised block design was followed in the selection of the samples, with IQ as the blocking variables. The samples consisted of three groups of size 32 each composed of students of standard IX selected from Tamil Nadu State Board schools covering one rural and two urban. The underachievers in the sample were identified by using the regression analysis. The tools used included CAI software on the language of sets, achievement test, and cultural fair, intelligent test by Cattell and Cattell, study habits inventory by Patel, and Mathematics study attitude scale by Sundarrajan, Mean, S.D, t-test, Chi-square, one-way and two-way ANOVA were used to treat the collected data. Major findings (1) Both the CAI strategies were superior to the traditional method of instruction, and CAI with TSS was more effective than CAI without TSS for underachiever (UA). (2) Except achievement level, all the other learner variables combined with the treatment had no interaction effect on the achievement score.

(3) There was no relationship between the post-treatment scores and the variables 'sex', 'locali' and 'achievement level 'of the experimental group. In the case of the variables IQ, 'Study habits' and 'maths study attitude', the positive relationship between those variable and achievement at the pre-treatment level was found to be cancelled at the post-test. Similar results were obtained for UA.

(Pilli, 2008) conducted a study on The Effects of Computer-Assisted Instruction on The Achievement, Attitudes and Retention of Fourth Grade Mathematics Course. A Ph.D. thesis submitted to Middle East Technical University. The purpose of this study was to examine the impact of computer assisted instruction with the software Frizbi Mathematics 4 on fourth grade students' achievement, attitudes and retention in mathematics lessons. In this study quasi-experimental research design was used in order to investigate the impacts of the Frizbi Mathematics 4 educational software on the 4th grade student's mathematics achievement, mathematics attitude, and computer assisted learning attitude, and retention. Research Questions were 1: Is there a significant difference between the achievement post-tests scores of the students exposed to Computer Assisted Instruction with the Frizbi Mathematics 4 and those who were exposed to traditional instruction with textbook? 2: Is there a significant difference between the mathematics attitude scale post scores of the students exposed to computer assisted instruction with Frizbi Mathematics 4 and those who were exposed to traditional instruction with textbook? 3: Is there a significant difference between the computer attitude scale post scores of the students exposed to computer with Frizbi Mathematics 4 and those who were exposed to traditional instruction with textbook? 4: Is there a significant difference between the retention test scores of the students exposed to computer assisted instruction with the Frizbi Mathematics 4 and those who were exposed to traditional instruction with textbook? The sample consisted of 26 students in control group and 29 students in Experimental Group. Findings of the study 1. The results of pre-test and post-tests for unit 1: "Multiplication of Natural Numbers" revealed that the CAI with

Frizbi Mathematics 4 applied to the experimental group was demonstrated to be effective in increasing the 115 achievement scores of the students. 2. The results of pre-test and post-tests for unit 2: "Division of Natural Numbers" were significant differences between achievement tests' mean scores of students in the experimental and control group. 3. The results of pre-test and post-tests for unit 3: "Fractions" revealed there were significant differences between achievement tests' mean scores of students in the experimental and control group. 4. In unit 1 (Multiplication of Natural Numbers) retention test were lower than the post-test mean scores in both groups and the "rate of retention decay" was not significantly different between the experimental and the control group, the results of independent t-test indicated that the experimental group's retention test mean score was significantly higher than the control group.

(Jackson & Dave, 2011) conducted a study on "The Effect of Computer-Assisted Instruction on Student's Attitudes and Achievement in Matrices and Transformations in Secondary Schools in Uasin Gishu District, Kenya, Moi University, Kenya". The purpose of the study was to investigate the effects of CAI on students' attitude and achievement in matrices and transformations between form four students who received instruction using CAI module or conventional instruction methods. The study addressed the following questions: 1. What are the effects of the CAI module on students' achievement in matrices and transformations? 2. Is there any significant difference in the achievement on matrices and transformations between subjects exposed to CAI module and those not? 3. What are the effects of the CAI module on students' attitudes towards Mathematics course? 4. Is there any significant difference in attitudes towards lessons on matrices and transformations between subjects exposed to CAI module and those not? The pre-test – post-test control group experimental research design was used. Six classes selected at random with 205 students participated in the study. Results of this study indicated higher achievement and positive attitudes with CAI treatment groups. Making connections between the goals of Mathematics education and CAI offers a valuable means for improving mathematical knowledge and skills and hence performance in Mathematics.

(Bayturan & Kesan, 2012) conducted a study on "The Effect of Computer Assisted Instruction on the Achievement and Attitudes towards Mathematics of Students in Mathematics Education". The objective of this study was to investigate the impact of Computer Assisted Instruction method on students achievement and attitudes towards mathematics in secondary mathematics education. The research was designed based on an experimental pre-test post-test model. The research was conducted in 60 ninth grade students from a Anatolian high-school during 2009-2010 academic year. The experiment group consists of 30 students and the control group consists of 30 students. The research is implemented by using Computer Assisted teaching material that is developed by Flash MX program related with the unit of "Relation, Function and Operation" of the area of learning algebra and took 10 weeks. Computer Assisted Instruction and traditional instruction methods were used in the experiment group and the control group respectively. The data were collected by using the Mathematics Test, Mathematics Attitudes Scale. Data analysis was done using t-test. The results demonstrated that teaching mathematics with a computer assisted instruction method increased student success significantly in mathematics lesson. However, the experimental and control groups did not differ between students' attitudes towards mathematics.

(Pramila, 2014) conducted a study entitled, "Development and Implementation of Computer Assisted Instruction in Mathematics for Standard VIII Students". Objectives of this study were (1) To develop the CAI in Mathematics for Standard VIII GSHSEB (Gujarat State Secondary and Higher Secondary Education Board) students. (2) To study the effectiveness of the developed CAI in terms of students' achievement in Mathematics with one of the experimental groups of standard VIII students. (3) To study the effectiveness of the developed CAI in terms of students' achievement in Mathematics with another experimental group of standard VIII students along with treatment of simultaneous discussion. (4) To study the relative effectiveness of learning mathematics in class VIII among the three groups A, B and C (Where C is the control group and A and B are experimental groups) in terms of achievement of the students. (5) To study the reaction of the students belonging to experimental groups about the mode of learning mathematics at the end of the experimentation. Study revealed that CAI with simultaneous discussion is more effective than the conventional method in initial and final try outs therefore CAI should be developed in the topics of mathematics wherever students find difficultly in learning.

5.2. The Related Literature in the Field of E-learning Other Than Mathematics

(Strother, 2002) conducted a study on "An Assessment of the Effectiveness of e-learning in Corporate Training Programs". The objective of the study were (1) measuring the results of e-learning through level 1 reaction, How do Learners Feel? How do e-Learning Instructors Feel? (2) Level II – Learning, What do Research Studies Show About Level II E-Learning? (3) Level III—Behavior, (4) Level IV – Results (5) Level V – ROI (return of investment) (5) Content Quality Measures. Findings of the study were Eighty-seven percent preferred to take digital courses during work hours. Fifty-two percent preferred e-learning in a workplace office area. Eighty-four percent would take a similar e-course if offered again. Thirty-eight percent said they generally preferred e-learning to classroom training. 62 percent said their students learned equally effectively in the online environment; 23 percent of faculty stated that their students learned better online; while 90 percent indicated that they were satisfied with online teaching. One faculty comment was: "Online students participate more, perform slightly better than, and are at least as satisfied as their on campus counterparts.

(Andrew R. Brown, 2005) conducted a study on, "Elements of Effective e-Learning Design". They designed phase of the e-learning resources. By 'design' they mean the planning or working out of the e-learning resource. This combines tasks including lesson planning, instructional design, creative writing, and software specification. Their design considerations take into account the constraints of building and delivery, but do not assume any particular software methodology or development platform. They found that in the TLF learning object Sonic Space City, where students create a soundscape to visual stimulus, feedback mechanisms include the ability to monitor work in progress in real time, and the facility to analyze and enhance projects by recording and replaying audio. In addition, students can describe their finished soundscape according to whether they consider it peaceful or noisy, calm or busy, scary or safe. These descriptions may be printed and the views of other class members sought. In this design, deliberate attention has been paid to incorporating multiple avenues for feedback to the learner, both during and after the activity. Students who use this learning object have displayed high levels of motivation and engagement because of the immediate and rich feedback provided by the activity. Activities such as this might be enhanced by enabling feedback from a broader range of people *via* the presentation of their soundscapes in an online forum, where reviews and downloads could provide qualitative and quantitative feedback about the appeal of their musical creation to others.

(Rita Birzina, 2009) conducted a study on "Effectiveness of interactive e-learning organization and quality assurance in European inter-university master studies." The sequential explorative qualitative and quantitative research design comprised of the following stages: exploration of the context of ICT use, qualitative and quantitative data processing of e-platform data, data analysis and interpretation. The sample consisted of 21 students of the Master's programme studying at the University of Latvia. Qualitative data processing was done applying AQUAD 6.0 software and quantitative data processing – applying SPSS 16.0 software. The study revealed that the use of ICT in the studies is effective if it provides student's personal experience, self-feeling and social recognition as conditions for creation of new knowledge, students' competencies and learning outcomes.

(Umek *et al.,* 2015) in their studies "Analysis of Selected Aspects of Students' Performance and Satisfaction in a Moodle-Based E-learning System Environment". The use of e-learning techniques in higher education is becoming ever more frequent. In some institutions, e-learning has completely replaced the traditional teaching methods, while in others it supplements classical courses. The paper presents a study conducted in a member institution of the University of Ljubljana that provides public administration programmes. We analysed the relationship between the proportion of the course implemented in the Moodle e-learning platform and students' performance on one hand and their satisfaction on the other. The empirical findings reveal a positive correlation for both elements. The results can help decision-makers learn more about how to enhance students' success and satisfaction using an e-learning platform.

(Anshari *et al*., 2015) in their studies "Pervasive Knowledge, Social Networks, and Cloud Computing: E-learning 2.0". Embedding Web 2.0 in learning processes has extended learning from traditional based learning-centred to a collaborative based learning-centred institution that emphasises learning anywhere and anytime. While deploying Semantic Web into e-learning offers a broader spectrum of pervasive knowledge acquisition to enrich users' experience in learning. In conventional learning practices, a student is perceived as a recipient of information and knowledge. However, nowadays many institutions involve students in learning processes that play an active role in improving collaborative learning platform and knowledge sharing as well as participation. For education provider, choosing a collaborative learning platform is becoming more important than choosing a tool in the new paradigm. Implementing collaborative learning through Web 2.0 is attributed to e-learning 2.0 as an approach for an online learning solution. Researchers have employed contents analysis for reviewing literatures in peer-reviewed journals. In fact, researchers found e-learning 2.0 is not only addressing the need redefining technologies like cloud computing in e-learning but also the need of redefining value added through pervasive knowledge generated from Semantic web and Social networks. E-learning 2.0 can be a solution to make learning systems in a higher education is flexible in terms of resources adoption, pervasive knowledge acquisition, and cloud computing implementation.

(Kenan, 2015) in their study titled "Improving the Effectiveness of E-learning Implementation in the School of Engineering at Tripoli University". The study is concerned with generating recommendations for the development of e-learning strategy at the School of Engineering at Tripoli University in Libya. They are based on the identified barriers to the successful e-learning implementation in the institution after the quantitative and qualitative analysis of the questionnaires developed using Smart Survey software package. The barriers were classified in three main categories: technical, cultural and mismanagement. These barriers were compared with those from Middle East and North Africa (MENA) countries (as close culture) and UK (as international culture and developed country). Then a SWOT (Strengths, Weaknesses, Opportunities, and Threats) model for the diagnostic of the current stage of e-learning performance in the institution is developed. The user-centred design and action research approach are used for the design, development and implementation of an e-learning package for a module studied by Year 4 students from School of Engineering at Tripoli University. The e-learning packages played a supporting role in the delivery of the chosen module therefore the blended learning approach contributed to the improvement of the quality for the teaching and learning processes. A novel I-CUBE model which can be used for the development of teaching and learning activities in a digital era is proposed. The sources of information for the proposed model are: quantitative and qualitative analysis of the stakeholders answers; conclusions for the (SWOT) analysis; researcher's reflections of the design, development and implementation of the e-learning package, study of the relevant educational publications and researcher's personal experience. There are several common opinions about the factors influencing the successful e-learning implementation so it is possible to make correlations between the stakeholders' points of view and to construct the surfaces which are included in the I-CUBE model. A set of institutional, pedagogical and technological recommendations for the development of e-learning strategy in the School of Engineering at Tripoli University are formulated. These are derived from the analysis of barriers and SWOT model related to the successful e-learning implementation in Libyan Higher Education Institutes (LHEIs). Also the aspects included in the proposed novel I-CUBE model for the development of teaching and learning activities, the researcher's personal experience as a student and lecturer at Tripoli University and as a Ph.D. student at the University of Huddersfield in UK are also considered. The implementation of these recommendations will enable the enhancement of students' learning experience and staff satisfaction through technology-based education addressing the flexible and diverse learning community's needs.

(Kiravu *et al.,* 2016) in their studies "E-learning Hands-On: Blending Interactive e-learning with Practical Engineering Laboratory". Purpose: Integrating laboratory work into interactive engineering e-learning contents augments theory with practice while simultaneously ameliorating the apparent theory-practice gap in traditional e-learning. The purpose of this paper is to assess and recommend media that currently fulfil this desirable dual pedagogical goal. Design/methodology/approach: The qualitative approach compares the e-learner-content interactivity deriving from Mathematica's Computable Document File (CDF) application, Pearson's myLab and Lucas-Nuelle's UniTrain-I. Illustrative interactive examples written in JavaScript and Java are thereby drawn from an engineering e-learning course developed at the University of Botswana (UB). Findings: Based on its scientific rigour, wide application scope, engineering analytical depth, minimal programming requirements and cross-subject-cum-faculty application and deployment potential, the authors found the CDF to be a versatile environment for generating dynamically interactive e-learning contents. The UniTrain-I, blending a multimedia information and communication technology (ICT)-based interactive e-learner-content philosophy with practical laboratory experimentation, is recommended for meeting the paper's dual e-learning goal as the most adept framework to-date, blending dynamic interactive e-learning content with laboratory hands-on engineering experimentation. Research limitations/ implications: The lack of other competing frameworks limited the considerations to only the three mentioned above. Consequently, the results are subject to review as the ongoing research advances new insights. Originality/value: The conclusions help e-learning designers plan ICT-based resources for integration into practical electrical engineering e-learning pedagogy and both CDF and UniTrain-I help dispel the prevailing apparent disquiet regarding the effectiveness of the e-learning-mediated electrical engineering pedagogy. In addition, the cited examples document an original electrical engineering e-learning course developed at the UB.

(Liu, Sheng, Hsueh & Sung-Lin, 2016) in their studies "Effects of Digital Teaching on the Thinking Styles and the Transfer of Learning of the Students in Department of Interior Design". Along with the constant advance of information technology and the rapid development of the Internet, the diverse functions and characteristics of e-learning break through lots of limitations in traditional instruction. Properly integrating e-learning design with Internet activities could enhance students' learning effect, and applying digital technology to assist in teachers' instruction is a future trend of instruction. A non-equivalent experimental design is applied to the quasi-experimental research in this study, in which total 162 students in three classes in National Taipei University of Technology, Tung Fang Design Institute, National Yunlin University of Science and Technology in Taiwan are selected as the research subjects for the 16-week (three hours per week) experimental teaching research. The research results conclude that 1. Thinking styles present significant effects on transfer of learning, 2. E-learning shows remarkable effects on transfer of learning, 3. Liberal thinking styles reveal the best effect on promoting far transfer under e-learning, and 4. Conservative thinking styles appear the best effect on promoting near transfer under e-learning. At the end, conclusions and suggestions are proposed in this study, expecting to provide teachers with some assistance in the teaching methods.

(Kiravu C., 2016) in their studies "e-learning hands-on: blending interactive e-learning with practical engineering laboratory". Integrating laboratory work into interactive engineering e-learning contents augments theory with practice while simultaneously ameliorating the apparent theory-practice gap in traditional e-learning. The purpose of this paper is to assess and recommend media that currently fulfil this desirable dual pedagogical goal. The qualitative approach compares the e-learner-content interactivity deriving from Mathematica's Computable Document File (CDF) application, Pearson's myLab and Lucas-Nuelle's UniTrain-I. Illustrative interactive examples written in JavaScript and Java are thereby drawn from an engineering e-learning course developed at the University of Botswana (UB). Based on its scientific rigour, wide application scope, engineering analytical depth, minimal programming requirements and cross-subject-cum-faculty application and deployment potential, the authors found the CDF to be a versatile environment for generating dynamically interactive e-learning contents. The UniTrain-I, blending a multimedia information and communication technology (ICT)-based interactive e-learner-content philosophy with practical laboratory experimentation, is recommended for meeting the paper's dual e-learning goal as the most adept framework to-date, blending dynamic interactive e-learning content with laboratory hands-on engineering experimentation. The lack of other competing frameworks limited the considerations to only the three mentioned above. Consequently, the results are subject to review as the ongoing research advances new insights. The conclusions help e-learning designers plan ICT-based resources for integration into practical electrical engineering e-learning pedagogy and both CDF and UniTrain-I help dispel the prevailing apparent disquiet regarding the effectiveness of the e-learning-mediated electrical engineering pedagogy. In addition, the cited examples document an original electrical engineering e-learning course developed at the UB.

(Kattoua, Al-Lozi & Alrowwad, 2016) in their studies "A Review of Literature on E-learning Systems in Higher Education". Due to the rapid growth of internet technology, universities around the world are investing heavily in e-learning systems to support their traditional teaching and to improve their students' learning experience and performance. However, the success of an e-learning system depends on the understanding of certain antecedent factors that influence the students' acceptance and usage of such e-learning systems. This study aims to provide a discussion of the current e-learning environments including their characteristics, limitations, advantages and the major factors that affect the acceptance of such technologies. It is concluded that a successful e-learning system should consider the personal, social, cultural, technological, organizational and environmental factors.

(Zare *et al.,* 2016) in their studies "Multi-criteria decision making approach in E-learning: A systematic review and classification". Academic research on E-learning has increased extensively over the past few years. Although, many multi-criteria decision making methods have been proposed to evaluate and examine the effectiveness of E-learning, there is a lack of study concerning systematic literature review and classification of research in this area. Regarding this, five major databases including Science Direct, Emerald, Taylor and Francis, IEEE, and Springer have been selected and a systematic methodology proposed. Consequently, a review of 42 published papers appearing in 33 academic journals and international conferences between 2001 and 2015 have been obtained to achieve a comprehensive review of MCDM application in E-learning. Accordingly, the selected papers have been classified by the year of publication, MCDM techniques, and journals and conferences in which they appeared. In addition, the significant criteria in evaluating E-learning were found. This study supports researchers and practitioners in effectively adopting MCDM techniques regarding E-learning evaluation and provides an insight into its state-of-the-art.

(Basogain, Olabe & Olabe, 2017) in their studies "Transition to a Modern Education System through e-learning". All around the world, governments are using international benchmarking of education systems as the method to identify the top performing countries. After the world best education systems have been identified, they are used as models for the rest of the countries. This procedure has an obvious limitation: if the rankings are obtained assessing content and methods already established, these are perpetuated. New curriculum and methodologies are difficult to integrate into existing education systems. Research in the last two decades has shown the limitations of the current education systems and the need to incorporate new paradigms. However, the current method that the governments use to design their education systems makes the adoption of radically new paradigms a highly risky proposition. This paper reviews the limitations of the current education systems, in particular in the area of mathematics. It discusses a set of fundamental changes in curriculum and teaching methodology. Finally, it reviews the role of e-learning as an integral part of the transition from traditional education systems to modern systems.

(Lubov *et al.*, 2017) in their studies "RLCP-Compatible Virtual Laboratories with 3D-Models and Demonstration Mode: Development and Application in e-learning". The article describes the features of the application of 3D-models in the design and development of RLCP-compatible virtual laboratories for interactive practical exercises in MOOC giving an example of laboratory devoted to the gradient descent method (backtracking line search) to define a multivariable function. Also features of application of the virtual laboratory in visualizer mode to preview and preliminary theoretical training of students are described. The article also shows the results of a study on pilot application of the developed virtual laboratory in practical training of the discipline "Methods of optimization". It is shown that the use of 3D-models and the implementation of the visualizer mode helps to achieve better learning outcomes (a 10% correct answers increase, significant reduction of the execution time and number of the attempts at the assessment).

(Sheshasaayee & Bee, 2017) in their studies "E-learning: Mode to Improve the Quality of Educational System". Data mining plays a very important role in uncovering unseen information from large volumes of data. Educational data mining (EDM) is a specific data mining field applied in discovering invisible knowledge from the educational database. In EDM, e-learning environment is an important field to learn the context of learning environments. The intend of the research paper is to look up into the current trends in the e-learning environment. This paper mainly focuses, in understanding the research done and current trends in e-learning environment. This paper reports, how the previous scholars tackled the earlier/recent trends in e-learning research. This work makes use of Harzing's Publish software for analyzing and visualizing the conduct level of the research papers. This study utilizes cites space, for analyzing tendency in e-learning research and to summarize the learning model and design for educational technology system appropriately.

(Ren Lin & Kao, 2018) in their studies "Mental effort detection using EEG data in E-learning contexts". E-learning becomes an alternative learning mode since the prevalence of the Internet. Especially, the advance of MOOC (Massive Open Online Course) technology enables a course to enroll tens of thousands of online learners. How to improve learners' online learning experiences on MOOC platforms becomes a crucial task for platform providers. In this research, based on Cognitive Load Theory, we built a system to capture and tag a user's mental states while s/he is watching online videos with a commercial EEG device, and used different normalization schemes and time window lengths to process EEG signals recorded from the EEG device. Finally, we adopted different supervised learning algorithms to train and test the classifiers, and then evaluated their classification performance. The results show that the proposed approach can effectively process EEG data to train classifiers, which achieve high accuracy, precision and recall rates compared with those of previous studies. This system can effectively facilitate users' self-awareness of mental efforts in online learning contexts to enable the automatic feedback in synchronous and asynchronous learning contexts, especially taking MOOCs as an example.

(Garcia, Falkner & Vivian, 2018) in their studies "Self-Regulated Learning strategies using e-learning tools for Computer Science". In 1986, Barry Zimmerman and Manuel Martinez-Pons presented a taxonomy containing 14 categories on Self-Regulated Learning (SRL) strategies performed by high school students when studying. Since this study, researchers have used the taxonomy as a framework for their research on students' SRL strategies and behaviours. When the taxonomy was constructed in the mid-1980s, these categories did not consider students' SRL behaviours while using digital technologies to study. The goals of this systematic literature review are to understand if the original SRL strategies are addressed in modern Computer Science e-learning tools and to determine if tools have emerged that support SRL strategies not captured by the original taxonomy. This review organises the e-learning tools within the SRL categories. It shows that a preponderance of research has been done on certain SRL skills, with limited focus on other categories. The systematic literature review concludes with suggestions for future research opportunities pertaining to tools that support the original categories, as well as tools that support SRL strategies.

(Vallejo *et al.,* 2018) in their studies "Effect of Motivational Scaffolding on E-learning Environments: Self-Efficacy, Learning Achievement, and Cognitive Style". The present research studies the effects of motivational scaffolding that favor self-efficacy and improve learning achievement in students with different cognitive styles in the Field Dependence/Independence (FDI) dimension, when they interact in an e-learning environment on mathematics. The research has an experimental design with two groups and a posttest. One group of students interacted with an e-learning environment that contained the motivational scaffolding within its structure and the other group interacted with a computational environment without the scaffolding. The results showed significant differences in learning achievement and academic self-efficacy attributable to the effect of the scaffolding. In addition, it was found that the interaction with the computational environment neutralized cognitive style effects.

5.3. The Related Literature in the Field of E-learning in Mathematics

(Babor, Fassi, Hariri & Bideq, 2016) in their studies "An e-learning environment for algorithmic: Toward an active construction of skills". Assimilating an algorithmic course is a persistent problem for many undergraduate students. The major problem faced by students is the lack of problem solving ability and flexibility. Therefore, students are generally passive, unmotivated and unable to mobilize all the acquired knowledge (loops, test, variables, etc.) to deal with new encountered problems. Our study is structured around building, step by step, problem solving skills among novice learners. Our approach is based on the use of problem based learning in an e-learning environment. We begin by establishing a cognitive model which represents knowledge elements, grouped into categories of skills, judged necessary to be appropriated. We then propose a problem built on a concrete situation which aims to actively construct a skill category. We conclude by presenting around the proposed problem a pedagogical scenario for the set of learning activities designed to be incorporated in an E-learning platform.

(Borba, C. *et al.,* 2016) in their studies "Blended learning, e-learning and mobile learning in mathematics education". In this literature survey we focus on identifying recent advances in research on digital technology in the field of mathematics education. To conduct the survey we have used internet search engines with keywords related to mathematics education and digital technology and have reviewed some of the main international journals, including the ones in Portuguese and Spanish. We identify five sub-areas of research, important trends of development, and illustrate them using case studies: mobile technologies, massive open online courses (MOOCs), digital libraries and designing learning objects, collaborative learning using digital technology, and teacher training using blended learning. These examples of case studies may help the reader to understand how recent developments in this area of research have evolved in the last few years. We conclude the report discussing some of the implications that these digital technologies may have for mathematics education research and practice as well as making some recommendations for future research in this area.

(Bray & Tangney, 2017) in their study titled "Technology usage in mathematics education research a systematic review of recent trends". There is a significant body of research relating to technology-enhanced mathematics education and the perceived potential of digital tools to enhance the learning experience. The aim of this research is to take a structured look at the types of empirical interventions ongoing in the field, and to attempt to classify and analyse the ways in which digital tools are being employed in such research. A systematic analysis of 139 recent, published studies of technology interventions in mathematics education, selected from in excess of 2000 potential studies, has been undertaken. A system of classification, developed as a part of this research, is used to categorise the digital tools, the pedagogical foundations and goals of the activities, and the levels of technology integration in the studies. Analysis of the results of this classification highlights a disparity between what is being researched in published empirical studies, and approaches that have been recognised as optimising the potential of technology to enhance mathematics education. Potential reasons for current trends are proposed and explored. A classification system for technology interventions in mathematics education was developed. 139 studies were classified by Technology, Learning Theory, Aim, and level of integration (SAMR). Analysis of Results indicate high levels of usage of technology to augment traditional practice. Lower levels of transformative usage of technology were identified within the empirical research. Sustained, research-based support is required to achieve the full potential of technology.

(Miyazaki, 2018) in his studies "Web-based Application to Aid the Learning of Mathematical Proofs Using Logical Diagrams and Its Potential". The author developed a web-based application tool to help create mathematical proofs using logical diagrams in a tabular format. In the proposed method, we first verified the usefulness of the diagrammatization proposed in our previous study, and then proposed a web-based application tool that visualizes the process of applying inferences in proofs. Teachers and learners are asked to fill in relevant items into a tabular format to create diagrams illustrating mathematical proofs. This application automatically produces a diagram after a table is filled in with information on propositions, comments, and their relations (links). Another attempt for the updated version in the future is the implementation of GUI-based web application that can also store user logs including mouse trajectories to analyze their behaviors.

(Platz, Krieger, Niehaus & Winter, 2018) in their studies "Suggestion of an E-proof Environment in Mathematics Education". This paper deals with electronic proofs (e-proofs) implemented with existing open-source web applications (*i.e.,* IMathAS) to be used in school and academic teacher education in mathematics. In these e-proofs, the learner will be supported and guided to prove a given theorem. An educational e-proof environment is a preliminary step towards a complete proof with paper and pencil in order to support students in building arguments and overcoming difficulties related to a traditional approach. IMathAS is a web-based mathematics assessment tool for delivery and semi-automatic grading of mathematical homework. In this paper, a theoretical framework to identify preferences of learners for logical reasoning is presented. This framework determines the structure of the IT-environment. Therefore, a brief introduction to the system is given; benefits, requirements, and constraints as well as further perspectives are discussed.

5.4. The Related Literature in the Field of Problems in Learning Discrete Mathematics

(Junaedi, Suyitno, Sugiharti & Eng, 2015) in their studies titled "Disclosure Causes of Students Error in Resolving Discrete Mathematics Problems Based on NEA as A Means of Enhancing Creativity". This article is based on research cooperation of Foreign Affairs in the first year, which was carried out between the team of lecturers from the Department of Mathematics Education Unnes (Indonesia) with Dr. Chin Kin Eng, a lecture of Mathematics Education from University Malaysia Sabah. The main objective of this research in the first year is to uncover the cause of the students error in resolving Discrete Mathematics by Newman Errors Analysis (NEA). The Results of this research will be used as the basis for the subsequent research that reveal the mathematical creativity of the students. Outcomes of this research is the publication of research results in the International Journal and seminar at the international level. As a qualitative research, data collection through an analysis of the results of tests, questionnaires, observations, and interviews. Data analysis are data reduction, exposure of data, synthesising the data, triangulation, and the inference/verification. According to research, known that students errors in resolving discrete mathematics caused by: (1) the student did not know the meaning of a symbol or an existing term in the problem (Reading Errors), (2) the student did not understand the meaning of the problem, namely the student fails to write what is known and what is being asked (Comprehension Errors), (3) students forgot a formula that will be used or strategy/procedure what to do (Transformation Errors), (4) students could not make the problem-solving algorithms in sequence and correctly (Process Skills Errors), (5) the student could not answer according to the question (Encoding Errors), and (6) the student could not translate well, especially about which was written in English (Language Errors). Errors caused by carelessness students (Careless Errors) was not found. After giving Learning Therapy for the provision to resolve a problem through the algorithm and the correct sequence, the causes of the error of some students could be minimized, so that the number of students who were experiencing errors could also be reduced. Results of this research became the basis for continued research in the second year, which will be revealed and developed the mathematical creativity of students with prepare advanced research instruments.

5.5. The Related Literature in the Field of E-learning/ Technology in Discrete Mathematics

(Escribano, Antonio, Carbajo & Rosa, 2008) studied about "Interactive tools for Discrete Mathematics E-learning Education is a dynamical and complex area". Research in education increases year by year. This increase is an important factor in the development of new educational techniques. Nowadays, the information society in which we live needs wider educational spaces, and internet provides a powerful tool to introduce new educational technologies and to use online applications. Indeed, TICs (technologies for informatics and communication) provide a very positive aid to the learning tasks. Interactive tutorials allow access to a big amount of information in a multi-sequential way, by using specifically designed Java applets suitable for the learning of a specific knowledge. Due to these facts, the use of these tools has many advantages for learning mathematics. In this sense, we present here a hypertext with many interactive java applets, developed to show the main applications of Integer and Modular Arithmetic in a brief theoretical environment.

(Campit, Cayabyab & Galas, 2015) in their study titled "The Effect of Peer Tutoring on Achievement of Students in Discrete Structures". Mathematics is a difficult subject, however, the researcher believes that Mathematics teaching can be made easy depending upon several factors like the teaching strategy in Mathematics instruction. This study investigated the effect of peer tutoring strategy on the students' achievement in Discrete Structures of second year college students at the Pangasinan State University, Bayambang Campus, Philippines. More specifically, it attempted to determine and compare the achievements of the students exposed to peer tutoring and conventional teaching. The experimental method of research was used in the study particularly the posttest control group design. Two intact classes were randomly assigned to experimental and control group and were subjected to six weeks treatment period. Valid and reliable researcher-made achievement test was used as the data gathering instrument. The mean, standard deviation, skewness and analysis of covariance were computed to determine and compare the achievement of the two groups. The result of the study revealed that the students exposed to peer tutoring significantly achieved better than those exposed to conventional teaching. Based on the finding, it is concluded that the achievement of the students is enhanced when exposed to peer tutoring strategy. It is recommended that peer tutoring strategy should be used by mathematics teachers in order to improve the performance of their students in the subject. Furthermore, further studies should be conducted to find out the effect of peer tutoring in other subject areas.

(Kularbphettong, Putglan, Tachpetpaiboon, Tongsiri & Roonrakwit, 2015) in their studies titled "Developing of m-learning for Discrete Mathematics based on Android Platform". In this paper, we present Discrete Mathematics Learning system based on Android operating system. The objective of this system is to advice student to learn discrete mathematics more convenient on mobile platform. The design approaches and functional components of this system were described and this application was developed on Knowledge. In this project, it was divided the result by the research purposes into 2 parts: developing the Mobile application for students and testing and evaluating the system. Black box technique was used to evaluate application performances and Questionnaires were applied to measure user satisfaction with system usability by experts and students.

(Kamèeva *et al.,* 2016) in their studies "Adaptive Model for Discrete Math of the E-learning System". This paper deals with the principles of specification, modeling and implementation of an educational information system. The behavior in a sample e-learning application is described using a formal model, based on Petri Nets. An adaptive model of student's transitions during the course of an e-learning cycle is proposed and a comparison with non-adaptive systems is made. Depending on the type of student, learning style, level of knowledge and dynamic behavior, a general conclusion is drawn that the students spend more time in non-adaptive *vs*. adaptive e-learning systems.

(Hernández & González-Ramírez, 2017) in their studies "Design and assessment of the impact of an e-textbook in the engagement towards the learning of Discrete Mathematics". Mathematics is vital for the training of professionals from various branches. Current research shows that the number of students with lack of academic achievement in mathematics is significant. In the case of Discrete Mathematics, it learning increases its difficulty due to various motivational, attitudinal and ability variables. Hence, the need of promoting academic proposals that raises the levels of engagement of students towards the learning of the subject.

The present research project aims to assess the effects of an e-textbook on the engagement towards the learning of Discrete Mathematics in university students. A quasi-experimental quantitative methodology is used. It will begin in a first phase with the application of a questionnaire to measure the levels of engagement of students with the learning of Discrete Mathematics. Likewise, the levels of satisfaction of students and teachers with the study materials of Discrete Mathematic will be measured. A second phase will propose the design of an e-textbook as educational material. The results of the research will show the impact of the e-textbook elaborated in the engagement towards the learning of Discrete Mathematics, by a questionnaire applied to students who used the developed material. The expected results will widely allow the design of educational materials to enhance the engagement of other issues in different teaching scenario, which will be useful to improve the academic performance of the students in those subjects where the results are not the desired ones and it affects, therefore, the university quality standards.

(Ivanov & Saltan, 2017) In their paper "Discrete mathematics course supported by CAS MATHEMATICA", they discussed examples of assignments for a course in discrete mathematics for undergraduate students majoring in business informatics. We consider several problems with computer-based solutions and discuss general strategies for using computers in teaching mathematics and its applications. In order to evaluate the effectiveness of our approach, we conducted an anonymous survey. The results of the survey provide evidence that our approach contributes to high outcomes and aligns with the course aims and objectives.

(Blanco, 2018) In his paper, "Active Learning in a Discrete Mathematics Class" he described the active learning and collaborative learning activities implemented in an introductory mid-size discrete mathematics course for Informatics majors. Active learning and collaborative learning have been used to increase student engagement, but incorporating them in smaller classes is a completely different experience from doing so in larger classes. We offer some tips and suggestions on how to incorporate these activities in larger classes including the utilization of undergraduate teaching assistants during lectures, and allowing students to work together on worksheets during lectures with the help of the teaching staff. Course questionnaires collected from five different sections that ran in the spring, summer, and fall of 2016 with around 60 to 70 students suggest that this approach has been well-received. Furthermore, the DFW rate (the proportion of students that received a D, F, or withdrew from the class) of these sections was lower than the DFW rate of other sections that shared the same evaluations (exams, homework assignments, and quizzes) and grading scheme to determine the final letter grade.

5.6. The Related Literature in the Field of Barriers in Implementing E-learning in Education

(Rohayani, Kurniabudi & Sharipuddin, 2015) conducted a research entitled "A Literature Review: Readiness Factors to measuring e-learning Readiness in Higher Education" E-Readiness is recognizing as one the most critical aspect for achieving successfully implementation of E-learning programs in higher education. Understand the role of this factor could help university management to implement effective and efficient E-learning project. The Purpose of the study is to discuss the theory of E-learning Readiness Factors and investigate the readiness factors that have been found by previous researchers for measure e-readiness at higher education. This study applies literature review and meta-analysis as research methodology. Literature review conducts by review the relevant articles with topic "e-learning readiness". The study found that skills and attitudes are the most significant factors influences E-learning readiness.

(Deshmukh & Deshmukh, 2017) in their study titled "Analysis of various Barriers to implementation of Web based Technology in education: Literature Review". In recent years, e-learning is in boom. It is the result of technology adoption in education. Slowly traditional teaching styles have to be transformed to technology-based teaching. So Web based learning is the best solution for it. But traditional/ chalk and talk / face to face teaching method is deeply rooted into the nerves of Indian education system. Technology adoption in teaching style is going to face many barriers in adoption. This paper tries to review the relevant literature to find out the various types of barriers in implementation of web-based education or web based technology in teaching learning method.

Section 3

6. RATIONALE OF THE STUDY

Technology is used in our daily life for many purposes. Students use technology for playing video games, text message, video chat, surfing internet etc., We can use this technology for teaching school subjects. We can use this technology in innovative ways like creating virtual learning environment especially for active and collaborative forms of learning. Different technologies can be used to create different types of learning activity, different types of learners and different pace of learning. Personalized instruction can be given to suite individual learners. Many countries have given priorities to e-learning. (Beetham, 2005) Priorities for e-learning research have never been so widely debated. In the United Kingdom, the main research funding councils have commissioned a consultation process to identify key issues for future support, under the direction of the Institute for Educational Technology (McAndrew *et al.,* 2004).

Researcher has reviewed some studies to know the effectiveness of CAI and e-learning. According to (Rose Antony Stela, 1992) finding both the CAI strategies were superior to the traditional method of instruction, and CAI with TSS was more effective than CAI without TSS for underachiever. From other studies also we can find similar findings. Therefore researcher is interested in finding the effectiveness of e-learning in discrete mathematics for university students.

7. CONCEPTUAL FRAMEWORK

Investigators developed e-learning material in discrete mathematics. E-learning is the recent teaching learning process which uses technology. This material will enhance learning and will aim at individualized instruction. Students will be active learner and will minimize the work load of faculty. Sometimes it becomes difficult to find mathematics faculty because supply is less and demand is more. Moreover students will learn their own pace and will learn when they are ready. Physical presence of teacher is minimized in this process. Students have natural aversion towards mathematics. Most of them have difficulty in learning mathematics. There are other difficulties like scarcity of class room and availability of faculty. This learning material will overcome these practical difficulties. Students can sit comfortable in a place wherever they fell comfortable and learn discrete mathematics syllabus in their own pace. Therefore time and place of learning can be fixed by the learner.

8. DELIMITATION OF THE STUDY

The present study is delimited to universities in Tamil Nadu and only Discrete Mathematics offered in department of Computer Science and Mathematics.

9. OPERATIONALIZATION OF TERMS

- ***E-learning*:** Technology enhanced learning (TEL) which will be a self-learning software package developed and on line learning materials used by the investigators after observation of mathematics classroom to understand the student's ability, potential, grasping power and other learning behavior.
- ***Achievement in Mathematics*:** The grades obtained by each student in the achievement test constructed by the investigators in Discrete Mathematics.
- ***Effectiveness*:** In the context of the present study effectiveness refers to relative increase in the scholastic achievement of the students in post-test compared to that of pretest and the reaction of the students in reaction scale will be considered.

10. EXPLANATION OF TERMS

- ***Reaction of Students*:** The belief of the students regarding the effectiveness of learning mode of discrete mathematics.

11. RESEARCH QUESTIONS

(i) In what ways do learners benefit from personalized/ individualized learning?

(ii) What aspects of the learning experience and environment should be personalized/individualized for maximum benefit to learners?

(iii) What skills do practitioners need to take advantage of these opportunities?

(iv) What is the cost of these different forms of personalization (*e.g.*, provision of multiple resources; staff and learner time)?

12. OBJECTIVES OF THE STUDY

- To develop e-learning in Discrete Mathematics.
- To study the effectiveness of the developed e-learning in terms of students' achievement in Mathematics.
- To study the reaction of the students about the mode of learning discrete mathematics at the end of the experimentation.

13. HYPOTHESES

H_0 : There will be no significant difference in the achievement scores of pretest and post-test.

H_1 : There will be significant difference in the achievement scores of pretest and post-test.

Section 4

14. RESEARCH METHODOLOGY

14.1. Population of the Study

All students studying in universities of Tamil Nadu in the department of mathematics and computer science constitute the population.

14.2. Sample of the Study

The sample for the study is selected purposively. All Students of Central University of Tamil Nadu in the department of Mathematics and Computer Science having discrete mathematics in odd semester during the year 2017 is considered. Central University of Tamil Nadu students studying first semester during the year 2017 was selected as experimental group who studied selected topics of discrete mathematics through developed e-learning material. Another college which was selected randomly and having required facility was taken as control group which consisted of forty students. Control group students studied the same topics of discrete mathematics in the usual traditional face to face method.

14.3. Design of the Study

The present study is an experimental study. The investigators followed pretest posttest control design. Time taken for the study was one semester.

Phase I: Selection of sample for the study

The investigators went through the discrete mathematics syllabus offered in Department of Mathematics and Computer Science in the Central University of Tamil Nadu. The common topics in Discrete Mathematics syllabus of Central University of Tamil Nadu and selected college was selected for experimentation. The **selected topics** were Permutations, Groups, Equivalence Relations and Recurrence Relations.

Phase II: Development of the CAI

a. Investigators observed students of the selected group to know the mathematical ability, grasping level, potential, and level of understanding and other relevant learning behavior of students in mathematics class.

b. E-learning was developed based on the needs of the students, content, keeping in mind mathematical ability and other relevant learning behaviors by the investigators. It was shown to the experts in the field of Mathematics Education and the experts were informed about the observation regarding level of the students and their mathematical ability. The investigators duly incorporated the suggestions given by them and finalized the e-learning material.

Phase III

A Mathematics Achievement Test which was considered as pretest was prepared by the investigators. The questions were prepared from prerequisites for the selected topics in discrete mathematics. This test was administered to both experimental group students and control group students and marks were recorded.

Phase IV: Implementation of the developed e-learning material

Experimental Group students learned the selected topics in discrete mathematics with the help of developed e-learning material. Students were allowed to clear their doubts whenever they faced during this treatment by chatting online in Skype or contacting them whenever necessary with the investigators. Mean time control group students learnt the same topics in the usual traditional face to face method and for the same period of time. Investigators contacted the faculty who taught control group students in frequent intervals and coordination was maintained. Investigators ensured that control group students and experimental group students learnt the same topics in same time and times taken to cover the topics were same. Both went parallel so that equivalence was maintained. Experimental Group students were given unit test (online) at the end of each topic to know their progress and the marks were recorded. In total four online unit tests were conducted. At the end of each topic Reaction Scale which was prepared for those specific topics were administered.

Phase V: To study the effectiveness

The investigators administered the Mathematics Achievement Test in the selected topics of Discrete Mathematics after completion of treatment on students. The scholastic achievement of students in pretest and post-test were compared to judge the effectiveness of the developed e-learning material. **ANCOVA** was used to test the hypothesis.

A reaction scale of the students towards the treatment they have received during experimentation was developed by the investigators in all the selected topics. It was administered to the group to know their reaction towards the treatment received by them during the experimentation. The responses to the reaction scale was analyzed further to judge the effectiveness of the treatment given to the experimental group.

14.4. Tools for Data Collection

The investigators constructed Mathematics Achievement Test for pretest (related topics in mathematics) and post-test (portions from selected topics of discrete mathematics). The achievement test was prepared and was finalized prior to experimentation.

A reaction scale in all selected topics was developed by investigators to know the reaction of the students about the different treatments received by the experimental group.

14.5. Data Collection

The data collected for pretest and post-test was in interval scale and Reaction Scale was in ordinal scale. The required data was collected in terms of achievement scores with the help of achievement test conducted on group and responses given to the reaction scale by the students of experimental group formed the data to judge the effectiveness of the study.

14.6. Data Analysis

The collected data was analyzed quantitatively. Mathematics Achievement Scores was analyzed quantitatively using ANCOVA. Responses given by the students to the Reaction Scale of the experimental group was analyzed quantitatively using percentage.

14.6.1. *Data analysis of mathematics achievement score*

Data analysis was done using ANCOVA CALCULATOR Vassar stats online. The following are the table (Table 1, Table 2, Table 3, Table 4 and Table 5) for data analysis.

CV = concomitant variable DV = dependent variable

Table 1: ANCOVA entry

	Levels of independent variable			
	Sample A		***Sample B***	
Count	***CV***	***DV***	***CV***	***DV***
1	24	8	28	88
2	12	14	14	40
3	18	10	26	44
4	18	20	26	60
5	14	14	30	56
6	18	12	22	52
7	20	6	20	32

Table 1: (*Contd...*)

Table 1: *(Contd...)*

	Levels of independent variable			
	Sample A		*Sample B*	
Count	***CV***	***DV***	***CV***	***DV***
8	12	18	24	52
9	10	10	32	72
10	20	18	32	68
11	26	14	26	20
12	16	10	22	28
13	10	14	22	24
14	14	10	20	40
15	20	18	18	44
16	18	18	22	20
17	14	18	20	44
18	14	10	24	36
19	16	14	22	48
20	16	8	20	48
21	10	10		
22	12	16		
23	12	14		
24	20	12		
25	16	12		
26	16	12		
27	6	10		
28	20	22		
29	12	12		
30	10	10		
31	18	18		
32	18	10		
33	10	10		
34	8	10		
35	22	14		
36	26	16		
37	16	12		
38	16	16		
39	16	14		
40	18	12		

The following is the table (Table 2) for dependent variable.

Table 2: Dependent variable

	Sample	
A	B	Total
	n	
40	20	60
	Observed means	
13.15	45.8	24.0333
	Adjusted means	
15.2	41.69	24.03

Table 3: Result of correlation

Aggregate correlation within samples: CV vs DV	
r = 0.36	r2 = 0.13

Table 4: ANCOVA summary

Source	***SS***	***df***	***MS***	***F***	***P***
Adjusted means	5736.26	1	5736.26	58.86	<.0001
Adjusted error	5555.36	57	97.46		
Adjusted total	11291.62	58			

Table 5: Test for homogeneity of regressions.

Source	***SS***	***df***	***MS***	***F***	***P***
Between regressions	999.97	1	999.97	12.29	0.000904
Remainder	4555.39	56	81.35		
Adjusted error	5555.36	57			

Ancova calculator used: *http://vassarstats.net/vsancova.html*

From the above ANCOVA result it is clear that P <.0001 which implies that H_0 is rejected at .05 level of significance. Therefore, it is concluded that there is significance difference between mean marks of experimental group students and mean marks of control group students. It is observed from ANCOVA calculation that Adjusted Mean of Control Group student's marks is 15.2 and that of Experimental Group student's Adjusted Mean marks is 41.69. Mean marks of Experimental Group students is greater than Control Group students therefore we conclude that Experimental group students achievements who learnt through e- learning material is greater that achievements of Control Group students who learnt through face to face traditional method. **This showed the effectiveness of e-learning material developed by the investigators.** We can safely conclude that students' achievements in selected topics of Discrete Mathematics improved through the e-learning material and it enhances the learning of the learner.

Investigators continuously interacted with the Experimental Group students and cleared their doubts therefore e-learning material can be used as supplement to usual classroom teaching and **cannot replace teacher**. Moreover, experiment was conducted only for short period of time (one semester) and was not tested for longer period of time therefore we cannot say anything about long term learning using e-learning material.

14.6.2. *Data analysis of reaction scale*

Statement wise analysis of reaction scale.

Positive polarity statements are given points as follows in Table 6.

Table 6: Points for positive polarity statements.

Response	***Strongly agree***	***Agree***	***Not decided***	***Disagree***	***Strongly disagree***
Points	5	4	3	2	1

Negative polarity statements are given points as follows in Table 7.

Table 7: Points for negative polarity statements.

Response	***Strongly agree***	***Agree***	***Not decided***	***Disagree***	***Strongly disagree***
Points	5	4	3	2	1

Level of significance is .05 for analysis using Chi-square for all statements. Chi-square was calculated using online calculator (socscistatistics).

The Chi-square statistic compares the tallies or counts of categorical responses between two (or more) independent groups. Chi-square is a statistical test commonly used to compare observed data with data we would expect to obtain according to a specific hypothesis. Then we might want to know about the "goodness to fit" between the observed and expected. Were the deviations (differences between observed and expected) the result of chance, or were they due to other factors. How much deviation can occur before you, the investigator, must conclude that something other than chance is at work, causing the observed to differ from the expected? The chi-square test is always testing what scientists call the **null hypothesis,** which states that there is no significant difference between the expected and observed result.

Most common application for chi-squared is in comparing observed counts of particular cases to the expected counts.

We can calculate X^2:

$$X^2 = \frac{(x_1 - E_1)^2}{E_1} + \frac{(x_2 - E_2)^2}{E_2} + \ldots + \frac{(x_k - E_k)^2}{E_k} = \sum_{i=1}^{k} \frac{(x_i - E_i)^2}{E_i}$$

Data were analyzed through the statistical technique χ^2.

Statement 1: I like to learn Recurrence relations through e-learning.

***Polarity*:** Positive

Responses of the students in percentage is given in Table 8 and Fig. 1.

Table 8: Students response in percentage

Points	*Response*	*Response percent*	*Response count*
5	Strongly agree	10%	2
4	Agree	70%	14
3	Not decided	10%	2
2	Disagree	10%	2
1	Strongly disagree	0%	0

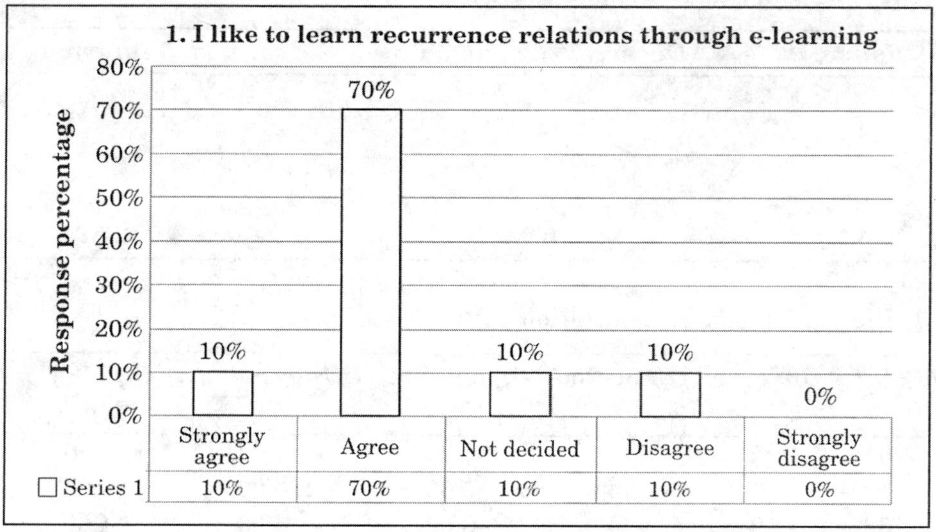

Fig. 1: Graphical representation of statement 1 in percentage.

10% of the students strongly agree with the statement "I like to learn Recurrence relations through e-learning".

70% of the students agree with the statement "I like to learn Recurrence relations through e-learning".

10% of the students not decided with the statement "I like to learn Recurrence relations through e-learning".

10% of the students disagree with the statement "I like to learn Recurrence relations through e-learning".

0.00% of the students strongly disagree with the statement "I like to learn Recurrence relations through e-learning".

Data analysis using Chi-square is given in Table 9 and Table 10.

H_0 : Response is uniformly distributed in the 5 point scale

H_1 : Response is not uniformly distributed in the 5 point scale

Table 9: Chi-square table for statement 1.

Points	*Observed frequency*	*Expected frequency*
5	2	4
4	14	4
3	2	4
2	2	4
1	0	4

Table 10: Chi-square calculation table for statement 1.

Points	*Observed*	*Expected*	*Difference*	*Difference sq.*	*Diff. Sq. / Exp fr.*
5	2	4	–2.00	4.00	1.00
4	14	4	10.00	100.00	25.00
3	2	4	–2.00	4.00	1.00
2	2	4	–2.00	4.00	1.00
1	0	4	–4.00	16.00	4.00
					32.000

The x^2 value is 32. The P-value is <0.001. The result is significant at $p = \leq 0.05$.

Table value of Chi-square at 4 df at 0.05 significance level is 9.49. Calculated value of Chi-square is more than the table value therefore, Null hypothesis is rejected. This revealed that reaction is not uniformly distributed in the 5 point scale. More load is on **agree** therefore most of the students **agreed** with the statement "I like to learn Recurrence relations through e-learning."

Statement 2: I can learn Recurrence relations without Teacher and only through e-learning material.

***Polarity*:** Positive

Responses of the students in percentage is given in Table 11 and Fig. 2.

Table 11: Students response in percentage.

Points	*Response*	*Response percent*	*Response count*
5	Strongly agree	0%	0
4	Agree	50%	10
3	Not decided	20%	4
2	disagree	25%	5
1	Strongly disagree	5%	1

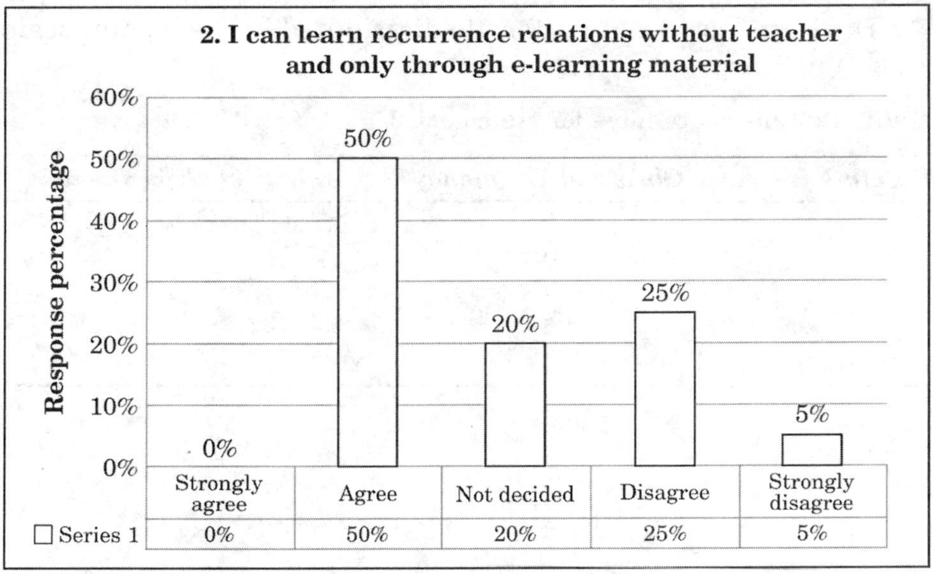

Fig. 2: Graphical representation of statement 2 in percentage.

0% of the students strongly agree with the statement "I can learn Recurrence relations without Teacher and only through e-learning material".

50% of the students agree with the statement "I can learn Recurrence relations without Teacher and only through e-learning material".

20% of the students not decided with the statement "I can learn Recurrence relations without Teacher and only through e-learning material".

25% of the students disagree with the statement "I can learn Recurrence relations without Teacher and only through e-learning material".

5% of the students strongly disagree with the statement "I can learn Recurrence relations without Teacher and only through e-learning material".

Data analysis using Chi-square is given in Table 12 and Table 13.

H_0 : Response is uniformly distributed in the 5 point scale

H_1 : Response is not uniformly distributed in the 5 point scale

Table 12: Chi-square table for statement 2.

Points	***Observed frequency***	***Expected frequency***
5	0	4
4	10	4
3	4	4
2	5	4
1	1	4

Table 13: Chi-square calculation table for statement 2.

Points	***Observed***	***Expected***	***Difference***	***Difference sq.***	***Diff. Sq. / Exp fr.***
5	0	4	–4.00	16.00	4.00
4	10	4	6.00	36.00	9.00
3	4	4	0.00	0.00	0.00
2	5	4	1.00	1.00	0.25
1	1	4	–3.00	9.00	2.25
					15.500

The x^2 value 15.5. The P-value<0.004. The result is significant at $p = \leq 0.05$

Table value of Chi-square at 4 df at 0.05 significance level is 9.49. Calculated value of Chi-square is more than the table value therefore, Null hypothesis is rejected. This revealed that reaction is not uniformly distributed in the 5-point scale. More load is on agree therefore most of the students agree with the statement “I can learn Recurrence relations without Teacher and only through e-learning material.”

Statement 3: I face problem in computer based learning at Recurrence relations.

***Polarity*:** Negative

Responses of the students in percentage is given in Table 14 and Fig. 3.

Table 14: Students response in percentage.

Points	***Response***	***Response percent***	***Response count***
1	Strongly agree	5%	1
2	Agree	35%	7
3	Not decided	20%	4
4	disagree	40%	8
5	Strongly disagree	0%	0

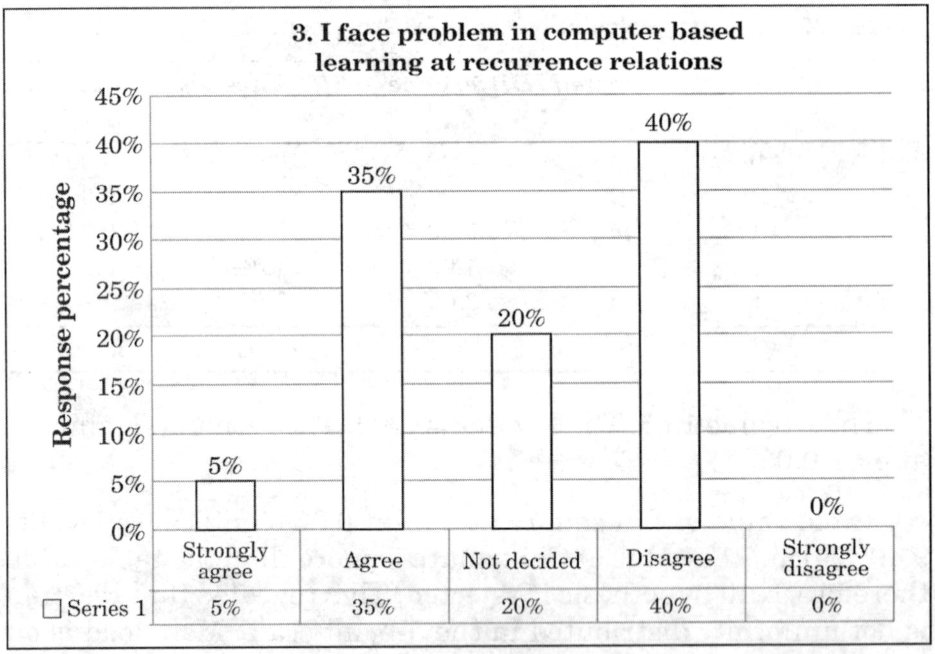

Fig. 3: Graphical representation of statement 3 in percentage.

5% of the students strongly agree with the statement "I face problem in computer based learning at Recurrence relations".

35% of the students agree with the statement "I face problem in computer based learning at Recurrence relations".

20% of the students not decided with the statement "I face problem in computer based learning at Recurrence relations".

40% of the students disagree with the statement "I face problem in computer based learning at Recurrence relations".

0% of the students strongly disagree with the statement "I face problem in computer based learning at Recurrence relations".

Data analysis using Chi-square are given in Table 15 and Table 16.

H_0 : Response is uniformly distributed in the 5 point scale

H_1 : Response is not uniformly distributed in the 5 point scale

Table 15: Chi-square table for statement 3.

Points	***Observed frequency***	***Expected frequency***
1	1	4
2	7	4
3	4	4
4	8	4
5	0	4

Table 16: Chi-square calculation table for statement 3.

Points	***Observed***	***Expected***	***Difference***	***Difference sq.***	***Diff. Sq. / Exp fr.***
1	1	4	–3.00	9.00	2.25
2	7	4	3.00	9.00	2.25
3	4	4	0.00	0.00	0.00
4	8	4	4.00	16.00	4.00
5	0	4	–4.00	16.00	4.00
					12.500

The x^2 value is 12.5. The P-value is <0.014. The result is significant at $p = \leq 0.05$

Table value of Chi-square at 4 df at 0.05 significance level is 9.49. Calculated value of Chi-square is more than the table value therefore, Null hypothesis is rejected. This revealed that reaction is not uniformly distributed in the 5 point scale. More load is on disagree therefore most of the students disagree with the statement "I face problem in computer based learning at Recurrence relations."

Statement 4: I face problem in answering Google's Quiz online test in Recurrence relations.

***Polarity*:** Negative.

Responses of the students in percentage is given in Table 17 and Fig. 4.

Table 17: Students response in percentage.

Points	*Response*	*Response percent*	*Response count*
1	Strongly agree	0%	0
2	Agree	35%	7
3	Not decided	15%	3
4	Disagree	50%	10
5	Strongly disagree	0%	0

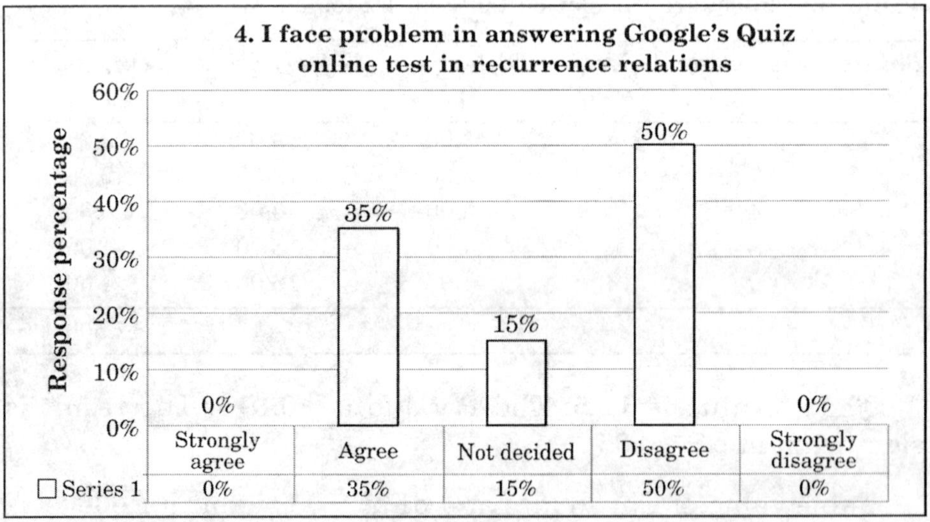

Fig. 4: Graphical representation of statement 4 in percentage.

0% of the students strongly agree with the statement "I face problem in answering Google's Quiz online test in Recurrence relations".

35% of the students agree with the statement "I face problem in answering Google's Quiz online test in Recurrence relations".

15% of the students not decided with the statement "I face problem in answering Google's Quiz online test in Recurrence relations".

50% of the students disagree with the statement "I face problem in answering Google's Quiz online test in Recurrence relations".

0% of the students strongly disagree with the statement I face problem in answering Google's Quiz online test in Recurrence relations".

Data analysis using Chi-square are given in Table 18 and Table 19.

H_0: Response is uniformly distributed in the 5 point scale

H_1: Response is not uniformly distributed in the 5 point scale

Table 18: Chi-square table for statement 4.

Points	*Observed frequency*	*Expected frequency*
1	0	4
2	7	4
3	3	4
4	10	4
5	0	4

Table 19: Chi-square calculation table for statement 4.

Points	*Observed*	*Expected*	*Difference*	*Difference sq.*	*Diff. Sq. / Exp fr.*
1	0	4	–4.00	16.00	4.00
2	7	4	3.00	9.00	2.25
3	3	4	–1.00	1.00	0.25
4	10	4	6.00	36.00	9.00
5	0	4	–4.00	16.00	4.00
					19.500

The x^2 value is 19.5. The P-value <0.001. The result is significant at $p = \leq 0.05$.

Table value of Chi-square at 4 df at 0.05 significance level is 9.49. Calculated value of Chi-square is more than the table value therefore, Null hypothesis is rejected. This revealed that reaction is not uniformly distributed in the 5-point scale. More load is on **disagree** therefore most of the students **disagreed** with the statement "**I face problem in computer based learning at Recurrence relations.**"

Statement 5: I face attention splash in e-learning for Recurrence relations.

Polarity: Negative

Responses of the students in percentage is given in Table 20 and Fig. 5.

Table 20: Students response in percentage.

Points	*Response*	*Response percent*	*Response count*
1	Strongly agree	5%	1
2	Agree	40%	8
3	Not decided	30%	6
4	Disagree	25%	5
5	Strongly disagree	0%	0

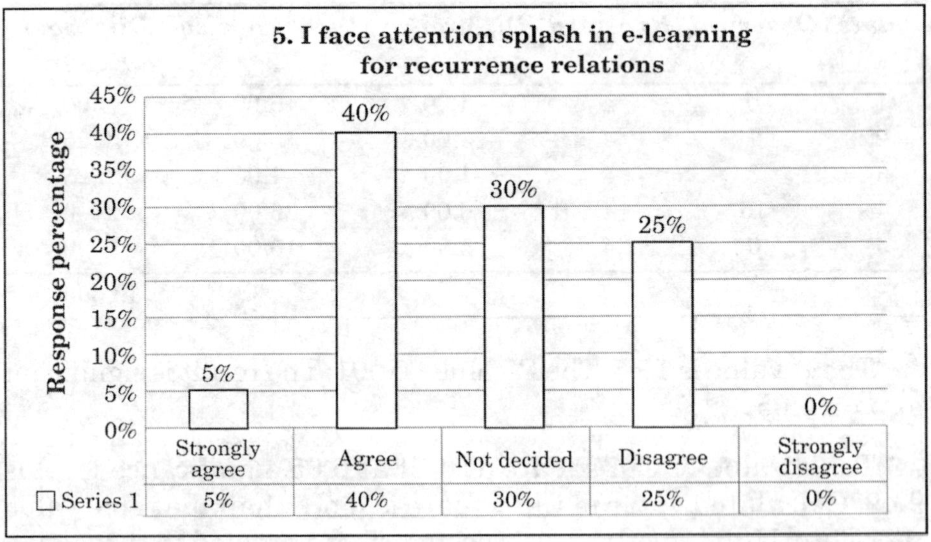

Fig. 5: Graphical representation of statement 5 in percentage.

5% of the students strongly agree with the statement "I face attention splash in e-learning for Recurrence relations".

40% of the students agree with the statement "I face attention splash in e-learning for Recurrence relations".

30% of the students not decided with the statement "I face attention splash in e-learning for Recurrence relations".

25% of the students disagree with the statement "I face attention splash in e-learning for Recurrence relations".

0.00% of the students strongly disagree with the statement "I face attention splash in e-learning for Recurrence relations".

Data analysis using Chi-square are given in Table 21 and Table 22.

H_0: Response is uniformly distributed in the 5 point scale

H_1: Response is not uniformly distributed in the 5 point scale

Table 21: Chi-square table for statement 5.

Points	*Observed frequency*	*Expected frequency*
1	1	4
2	8	4
3	6	4
4	5	4
5	0	4

Table 22: Chi-square calculation table for statement 5.

Points	*Observed*	*Expected*	*Difference*	*Difference sq.*	*Diff. Sq. / Exp fr.*
1	1	4	–3.00	9.00	2.25
2	8	4	4.00	16.00	4.00
3	6	4	2.00	4.00	1.00
4	5	4	1.00	1.00	0.25
5	0	4	–4.00	16.00	4.00
					11.500

The x^2 value is 11.5. The P-value is <0.021. The result is significant at $p = \leq 0.05$.

Table value of Chi-square at 4 df at 0.05 significance level is 9.49. Calculated value of Chi-square is more than the table value therefore, Null hypothesis is rejected. This revealed that reaction is not uniformly distributed in the 5 point scale. More load is on **agree** therefore most of the students **agreed** with the statement "I face attention splash in e-learning for Recurrence relations."

Statement 6: I know Education based websites for Recurrence relations.

***Polarity*:** Positive

Responses of the students in percentage is given in Table 23 and Fig. 6.

Table 23: Students response in percentage.

Points	*Response*	*Response percent*	*Response count*
5	Strongly agree	5%	1
4	Agree	85%	17
3	Not decided	5%	1
2	Disagree	5%	1
1	Strongly disagree	0%	0

5% of the students strongly agree with the statement "I know Education based websites for Recurrence relations".

85% of the students agree with the statement "I know Education based websites for Recurrence relations".

5% of the students not decided with the statement "I know Education based websites for Recurrence relations".

5% of the students disagree with the statement "I know Education based websites for Recurrence relations".

0.00% of the students strongly disagree with the statement "I know Education based websites for Recurrence relations".

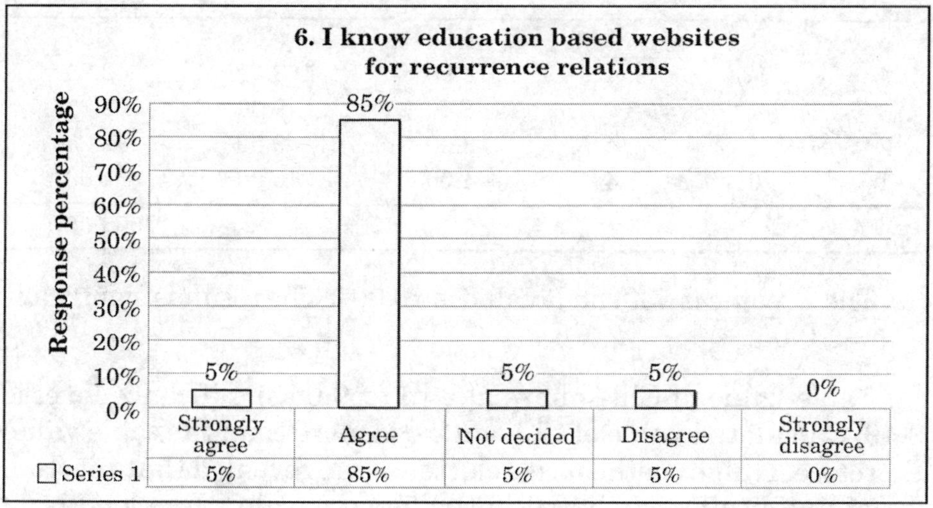

Fig. 6: Graphical representation of statement 6 in percentage.

Data analysis using Chi-square are given in Table 24 and Table 25.

H_0: Response is uniformly distributed in the 5 point scale

H_1: Response is not uniformly distributed in the 5 point scale

Table 24: Chi-square table for statement 6.

Points	***Observed frequency***	***Expected frequency***
5	1	4
4	17	4
3	1	4
2	1	4
1	0	4

Table 25: Chi-square calculation table for statement 6.

Points	*Observed*	*Expected*	*Difference*	*Difference sq.*	*Diff. Sq. / Exp fr.*
5	1	4	–3.00	9.00	2.25
4	17	4	13.00	169.00	42.25
3	1	4	–3.00	9.00	2.25
2	1	4	–3.00	9.00	2.25
1	0	4	–4.00	16.00	4.00
					53.000

The x^2 value is 53. The P-value is <0.001. The result is significant at $p = \leq 0.05$.

Table value of Chi-square at 4 df at 0.05 significance level is 9.49. Calculated value of Chi-square is more than the table value therefore, Null hypothesis is rejected. This revealed that reaction is not uniformly distributed in the 5 point scale. More load is on **agree** therefore most of the students **agreed** with the statement "I know Education based websites for Recurrence relations."

Statement 7: I face problem in learning Recurrence Relations through e-learning material.

***Polarity*:** Negative

Responses of the students in percentage is given in Table 26 and Fig. 7.

Table 26: Students response in percentage.

Points	*Response*	*Response percent*	*Response count*
1	Strongly agree	5%	1
2	Agree	35%	7
3	Not decided	5%	1
4	Disagree	55%	11
5	Strongly disagree	0%	0

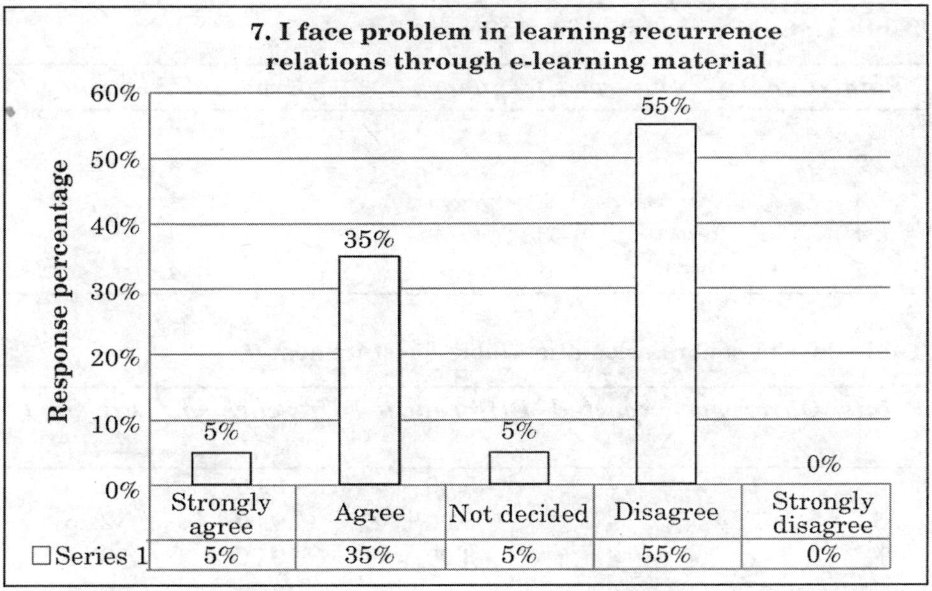

Fig. 7: Graphical representation of statement 7 in percentage.

5% of the students strongly agree with the statement "I face problem in learning Recurrence Relations through e-learning material".

35% of the students agree with the statement "I face problem in learning Recurrence Relations through e-learning material".

5% of the students not decided with the statement "I face problem in learning Recurrence Relations through e-learning material".

55% of the students disagree with the statement "I face problem in learning Recurrence Relations through e-learning material".

0.00% of the students strongly disagree with the statement "I face problem in learning Recurrence Relations through e-learning material".

Data analysis using Chi-square are given in Table 27 and Table 28.

H_0 : Response is uniformly distributed in the 5 point scale

H_1 : Response is not uniformly distributed in the 5 point scale

Table 27: Chi-square table for statement 7.

Points	*Observed frequency*	*Expected frequency*
1	1	4
2	7	4
3	1	4
4	11	4
5	0	4

Table 28: Chi-square calculation table for statement 7.

Points	*Observed*	*Expected*	*Difference*	*Difference sq.*	*Diff. Sq. / Exp fr.*
1	1	4	–3.00	9.00	2.25
2	7	4	3.00	9.00	2.25
3	1	4	–3.00	9.00	2.25
4	11	4	7.00	49.00	12.25
5	0	4	–4.00	16.00	4.00
					23.000

The x^2 value is 23. The P-value is <0.001. The result is significant at $p = \leq 0.05$

Table value of Chi-square at 4 df at 0.05 significance level is 9.49. Calculated value of Chi-square is more than the table value therefore, Null hypothesis is rejected. This revealed that reaction is not uniformly distributed in the 5 point scale. More load is on **disagree** therefore most of the students **disagreed** with the statement "I face problem in learning Recurrence Relations through e-learning material."

Statement 8: I get boredom in e-learning on learning Recurrence Relations.

***Polarity*:** Negative

Responses of the students in percentage is given in Table 29 and Fig. 8.

Table 29: Students response in percentage.

Points	*Response*	*Response percent*	*Response count*
1	Strongly agree	5%	1
2	Agree	30%	6
3	Not decided	25%	5
4	Disagree	40%	8
5	Strongly disagree	0%	0

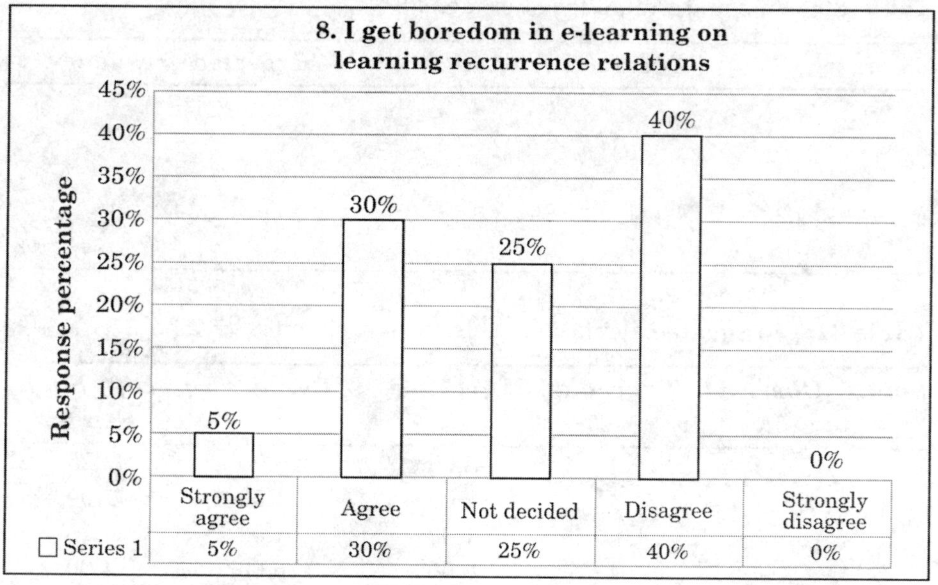

Fig. 8: Graphical representation of statement 8 in percentage.

5% of the students strongly agree with the statement “I get boredom in e-learning on learning Recurrence Relations”.

30% of the students agree with the statement “I get boredom in e-learning on learning Recurrence Relations”.

25% of the students not decided with the statement “I get boredom in e-learning on learning Recurrence Relations”.

40% of the students disagree with the statement “I get boredom in e-learning on learning Recurrence Relations”.

0.00% of the students strongly disagree with the statement "I get boredom in e-learning on learning Recurrence Relations".

Data analysis using Chi-square are given in Table 30 and Table 31.

H_0 : Response is uniformly distributed in the 5 point scale

H_1 : Response is not uniformly distributed in the 5 point scale

Table 30: Chi-square table for statement 8.

Points	*Observed frequency*	*Expected frequency*
1	1	4
2	6	4
3	5	4
4	8	4
5	0	4

Table 31: Chi-square calculation table for statement 8.

Points	*Observed*	*Expected*	*Difference*	*Difference sq.*	*Diff. Sq. / Exp fr.*
1	1	4	–3.00	9.00	2.25
2	6	4	2.00	4.00	1.00
3	5	4	1.00	1.00	0.25
4	8	4	4.00	16.00	4.00
5	0	4	–4.00	16.00	4.00
					11.500

The x^2 value is 11.5. The P-value is 0.021. The result is significant at $p = \leq 0.05$

Table value of Chi-square at 4 df at 0.05 significance level is 9.49. Calculated value of Chi-square is more than the table value therefore, Null hypothesis is rejected. This revealed that reaction is not uniformly distributed in the 5 point scale. More load is on **disagree** therefore most of the students **disagreed** with the statement "I get boredom in e-learning on learning Recurrence Relations".

Statement 9: I need teacher's help while learning recurrence relations through e-learning.

***Polarity*:** Negative

Responses of the students in percentage is given in Table 32 and Fig. 9.

Table 32: Students response in percentage.

Points	*Response*	*Response percent*	*Response count*
1	Strongly agree	30%	6
2	Agree	45%	9
3	Not decided	5%	1
4	Disagree	20%	4
5	Strongly disagree	0%	0

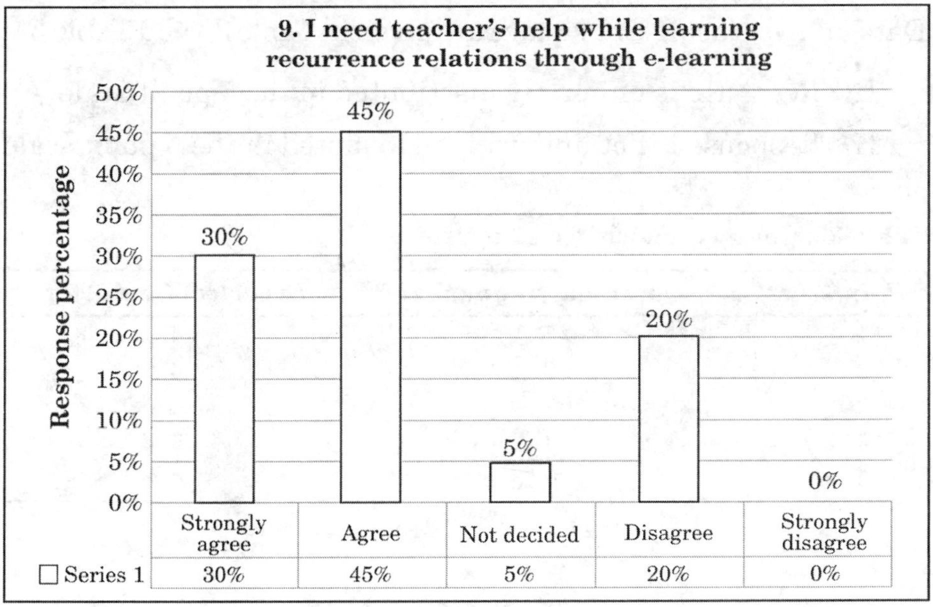

Fig. 9: Graphical representation of statement 9 in percentage.

30% of the students strongly agree with the statement “I need teacher’s help while learning Recurrence Relations through e-learning”.

45% of the students agree with the statement “I need teacher’s help while learning Recurrence Relations through e-learning”.

5% of the students not decided with the statement “I need teacher’s help while learning Recurrence Relations through e-learning”.

20% of the students disagree with the statement “I need teacher’s help while learning Recurrence Relations through e-learning”.

0.00% of the students strongly disagree with the statement “I need teacher’s help while learning Recurrence Relations through e-learning”.

Data analysis using Chi-square are given in Table 33 and Table 34.

H_0 : Response is uniformly distributed in the 5 point scale

H_1 : Response is not uniformly distributed in the 5 point scale

Table 33: Chi-square table for statement 9.

Points	*Observed frequency*	*Expected frequency*
1	6	4
2	9	4
3	1	4
4	4	4
5	0	4

Table 34: Chi-square table for calculation statement 9.

Points	Observed	Expected	Difference	Difference sq.	Diff. Sq. / Exp fr.
1	6	4	2.00	4.00	1.00
2	9	4	5.00	25.00	6.25
3	1	4	–3.00	9.00	2.25
4	4	4	0.00	0.00	0.00
5	0	4	–4.00	16.00	4.00
					13.500

The x^2 value is 13.5. The P-value is 0.009. The result is significant at $p = \leq 0.05$

Table value of Chi-square at 4 df at 0.05 significance level is 9.49. Calculated value of Chi-square is more than the table value therefore, Null hypothesis is rejected. This revealed that reaction is not uniformly distributed in the 5 point scale. More load is on **agree** therefore most of the students **agreed** with the statement "I need teacher's help while learning Recurrence Relations through e- learning".

Statement 10: I enjoyed e-learning as I can watch video in the topic recurrence relations.

***Polarity*:** Positive

Responses of the students in percentage is given in Table 35 and Fig. 10.

20% of the students strongly agree with the statement "I enjoyed e-learning as I can watch video in the topic Recurrence Relations".

65% of the students agree with the statement "I enjoyed e-learning as I can watch video in the topic Recurrence Relations".

10% of the students not decided with the statement "I enjoyed e-learning as I can watch video in the topic Recurrence Relations".

0% of the students disagree with the statement "I enjoyed e-learning as I can watch video in the topic Recurrence Relations".

5% of the students strongly disagree with the statement "I enjoyed e-learning as I can watch video in the topic Recurrence Relations".

Table 35: Students response in percentage.

Points	*Response*	*Response percent*	*Response count*
5	Strongly agree	20%	4
4	Agree	65%	13
3	Not decided	10%	2
2	Disagree	0%	0
1	Strongly disagree	5%	1

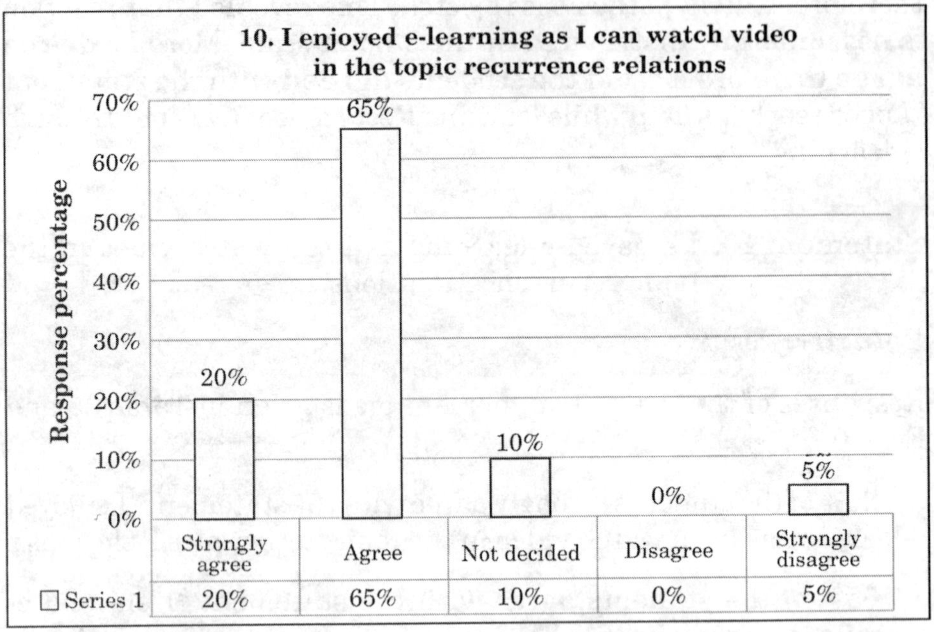

Fig. 10: Graphical representation of statement 10 in percentage.

Data analysis using Chi-square are given in Table 36 and Table 37.

H_0: Response is uniformly distributed in the 5 point scale

H_1: Response is not uniformly distributed in the 5 point scale

Table 36: Chi-square table for statement 10.

Points	*Observed frequency*	*Expected frequency*
5	4	4
4	13	4
3	2	4
2	0	4
1	1	4

Table 37: Chi-square calculation table for statement 10.

Points	*Observed*	*Expected*	*Difference*	*Difference sq.*	*Diff. Sq. / Exp fr.*
5	4	4	0.00	0.00	0.00
4	13	4	9.00	81.00	20.25
3	2	4	–2.00	4.00	1.00
2	0	4	–4.00	16.00	4.00
1	1	4	–3.00	9.00	2.25
					27.500

The x^2 value is 27.5. The P-value is 0.001. The result is significant at $p = \leq 0.05$

Table value of Chi-square at 4 df at 0.05 significance level is 9.49. Calculated value of Chi-square is more than the table value therefore, Null hypothesis is rejected. This revealed that reaction is not uniformly distributed in the 5-point scale. More load is on agree therefore most of the students agreed with the statement "I enjoyed e-learning as I can watch video in the topic Recurrence Relations".

Statement 11: I like to learn Permutations through e-learning.

***Polarity*:** Positive

Responses of the students in percentage is given in Table 38 and Fig. 11.

Table 38: Students response in percentage

Points	*Response*	*Response percent*	*Response count*
5	Strongly agree	15%	3
4	Agree	75%	15
3	Not decided	5%	1
2	Disagree	5%	1
1	Strongly disagree	0%	0

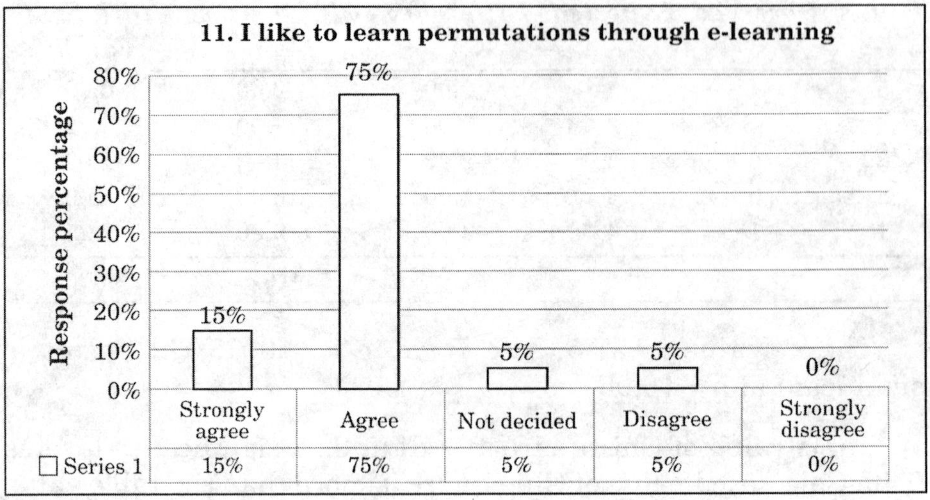

Fig. 11: Graphical representation of statement 11 in percentage.

15% of the students strongly agree with the statement "I like to learn Permutations through e-learning".

75% of the students agree with the statement "I like to learn Permutations through e-learning".

5% of the students not decided with the statement "I like to learn Permutations through e-learning".

5% of the students disagree with the statement "I like to learn Permutations through e-learning".

0.00% of the students strongly disagree with the statement "I like to learn Permutations through e-learning".

Data analysis using Chi-square are given in Table 39 and Table 40.

H_0 : Response is uniformly distributed in the 5 point scale

H_1 : Response is not uniformly distributed in the 5 point scale

Table 39: Chi-square table for statement 11.

Points	*Observed frequency*	*Expected frequency*
5	3	4
4	15	4
3	1	4
2	1	4
1	0	4

Table 40: Chi-square calculation table for statement 11.

Points	*Observed*	*Expected*	*Difference*	*Difference sq.*	*Diff. Sq. / Exp fr.*
5	3	4	–1.00	1.00	0.25
4	15	4	11.00	121.00	30.25
3	1	4	–3.00	9.00	2.25
2	1	4	–3.00	9.00	2.25
1	0	4	–4.00	16.00	4.00
					39.000

The x^2 value is 39. The P-value is 0.001. The result is significant at $p = \leq 0.05$

Table value of Chi-square at 4 df at 0.05 significance level is 9.49. Calculated value of Chi-square is more than the table value therefore, Null hypothesis is rejected. This revealed that reaction is not uniformly distributed in the 5 point scale. More load is on agree therefore most of the students agreed with the statement "I like to learn Permutations through e-learning".

Statement 12: I can learn Permutations without Teacher and only through e-learning material.

***Polarity*:** Positive

Responses of the students in percentage is given in Table 41 and Fig. 12.

Table 41: Students response in percentage.

Points	*Response*	*Response percent*	*Response count*
5	Strongly agree	15%	3
4	Agree	60%	12
3	Not decided	0%	0
2	Disagree	20%	4
1	Strongly disagree	5%	1

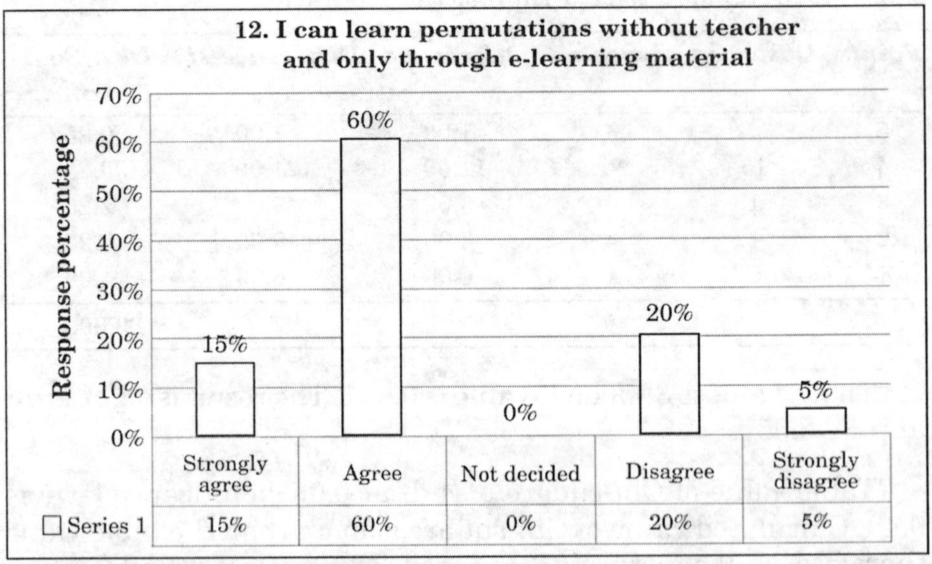

Fig. 12: Graphical representation of statement 12 in percentage.

15% of the students strongly agree with the statement "I can learn Permutations without Teacher and only through e-learning material".

60% of the students agree with the statement "I can learn Permutations without Teacher and only through e-learning material".

0% of the students not decided with the statement "I can learn Permutations without Teacher and only through e-learning material".

20% of the students disagree with the statement "I can learn Permutations without Teacher and only through e-learning material".

5% of the students strongly disagree with the statement "I can learn Permutations without Teacher and only through e-learning material".

Data analysis using Chi-square are given in Table 42 and Table 43.

H_0 : Response is uniformly distributed in the 5 point scale

H_1 : Response is not uniformly distributed in the 5 point scale

Table 42: Chi-square table for statement 12.

Points	*Observed frequency*	*Expected frequency*
5	3	4
4	12	4
3	0	4
2	4	4
1	1	4

Table 43: Chi-square calculation table for statement 12.

Points	*Observed*	*Expected*	*Difference*	*Difference sq.*	*Diff. Sq. / Exp fr.*
5	3	4	–1.00	1.00	0.25
4	12	4	8.00	64.00	16.00
3	0	4	–4.00	16.00	4.00
2	4	4	0.00	0.00	0.00
1	1	4	–3.00	9.00	2.25
					22.500

The x^2 value is 22.5. The P-value is 0.001. The result is significant at $p = \leq 0.05$

Table value of Chi-square at 4 df at 0.05 significance level is 9.49. Calculated value of Chi-square is more than the table value therefore, Null hypothesis is rejected. This revealed that reaction is not uniformly distributed in the 5 point scale. More load is on agree therefore most of the students agreed with the statement "I can learn Permutations without Teacher and only through e-learning material."

Statement 13: I face problem in computer based learning at Permutations.

Polarity: Negative

Responses of the students in percentage is given in Table 44 and Fig. 13.

Table 44: Students response in percentage.

Points	*Response*	*Response percent*	*Response count*
1	Strongly agree	5%	1
2	Agree	25%	5
3	Not decided	25%	5
4	Disagree	40%	8
5	Strongly disagree	5%	1

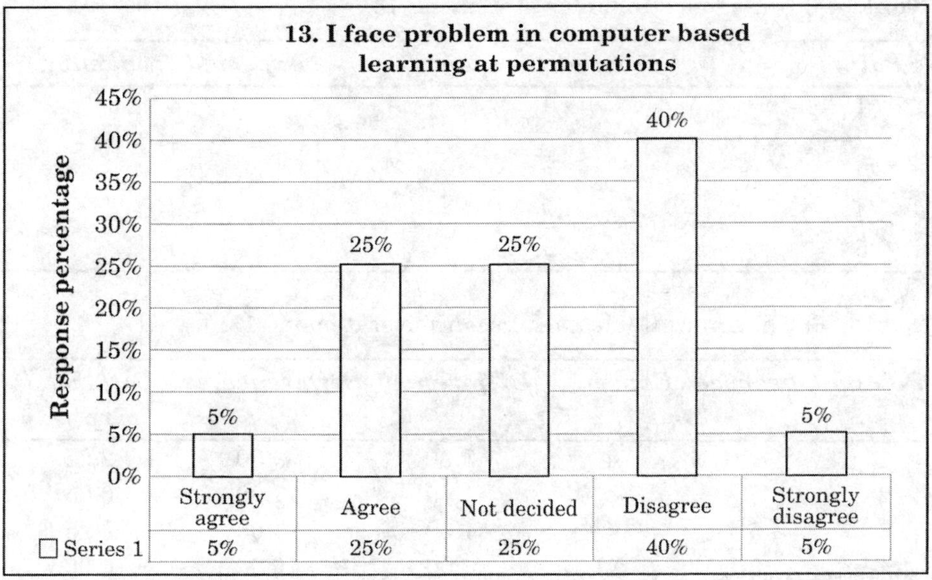

Fig. 13: Graphical representation of statement 13 in percentage.

5% of the students strongly agree with the statement "I face problem in computer based learning at Permutations".

25% of the students agree with the statement "I face problem in computer based learning at Permutations".

25% of the students not decided with the statement "I face problem in computer based learning at Permutations".

40% of the students disagree with the statement "I face problem in computer based learning at Permutations".

5% of the students strongly disagree with the statement "I face problem in computer based learning at Permutations".

Data analysis using Chi-square are given in Table 45 and Table 46.

H_0 : Response is uniformly distributed in the 5 point scale

H_1 : Response is not uniformly distributed in the 5 point scale

Table 45: Chi-square table for statement 13.

Points	*Observed frequency*	*Expected frequency*
1	1	4
2	5	4
3	5	4
4	8	4
5	1	4

Table 46: Chi-square table calculation for statement 13.

Points	*Observed*	*Expected*	*Difference*	*Difference sq.*	*Diff. Sq. / Exp fr.*
1	1	4	–3.00	9.00	2.25
2	5	4	1.00	1.00	0.25
3	5	4	1.00	1.00	0.25
4	8	4	4.00	16.00	4.00
5	1	4	–3.00	9.00	2.25
					9.000

The x^2 value is 9. The P-value is 0.061. The result is not significant at $p = \leq 0.05$.

Table value of Chi-square at 4 df at 0.05 significance level is 9.49. Calculated value of Chi-square is more than the table value therefore, Null hypothesis is rejected. This revealed that reaction is not uniformly distributed in the 5 point scale. More load is on disagree therefore most of the students disagreed with the statement "I face problem in computer based learning at Permutations".

Statement 14: I face problem in answering Google's Quiz online test in Permutations.

***Polarity*:** Negative

Responses of the students in percentage is given in Table 47 and Fig. 14.

Table 47: Students response in percentage.

Points	*Response*	*Response percent*	*Response count*
1	Strongly agree	5%	1
2	Agree	10%	2
3	Not decided	15%	3
4	Disagree	65%	13
5	Strongly disagree	5%	1

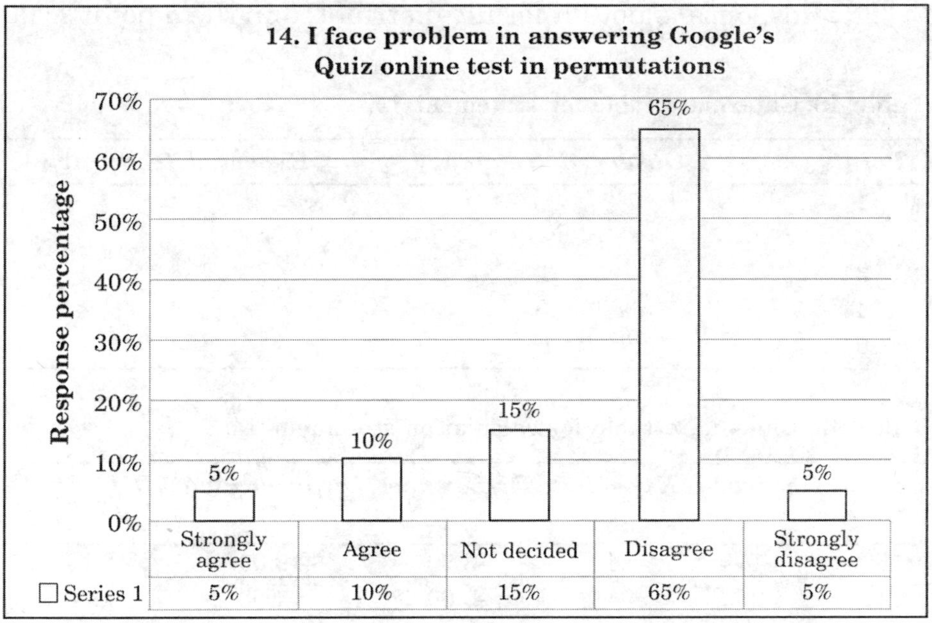

Fig. 14: Graphical representation of statement 14 in percentage.

5% of the students strongly agree with the statement "I face problem in answering Google's Quiz online test in Permutations".

10% of the students agree with the statement "I face problem in answering Google's Quiz online test in Permutations".

15% of the students not decided with the statement "I face problem in answering Google's Quiz online test in Permutations".

65% of the students disagree with the statement "I face problem in answering Google's Quiz online test in Permutations".

5% of the students strongly disagree with the statement "I face problem in answering Google's Quiz online test in Permutations".

Data analysis using Chi-square are given in Table 48 and Table 49.

H_0 : Response is uniformly distributed in the 5 point scale

H_1 : Response is not uniformly distributed in the 5 point scale

Table 48: Chi-square table for statement 14.

Points	*Observed frequency*	*Expected frequency*
1	1	4
2	2	4
3	3	4
4	13	4
5	1	4

Table 49: Chi-square table for calculation statement 14.

Points	*Observed*	*Expected*	*Difference*	*Difference sq.*	*Diff. Sq. / Exp fr.*
1	1	4	–3.00	9.00	2.25
2	2	4	–2.00	4.00	1.00
3	3	4	–1.00	1.00	0.25
4	13	4	9.00	81.00	20.25
5	1	4	–3.00	9.00	2.25
					26.000

The x^2 value is 26. The P-value is 0.001. The result is significant at $p = \leq 0.05$

Table value of Chi-square at 4 df at 0.05 significance level is 9.49. Calculated value of Chi-square is more than the table value therefore, Null hypothesis is rejected. This revealed that reaction is not uniformly distributed in the 5 point scale. More load is on disagree therefore most of the students disagreed with the statement "I face problem in answering Google's Quiz online test in Permutations".

Statement 15: I face attention splash in e-learning for Permutations.

***Polarity*:** Negative

Responses of the students in percentage is given in Table 50 and Fig. 15.

Table 50: Students response in percentage.

Points	*Response*	*Response percent*	*Response count*
1	Strongly agree	5%	1
2	Agree	35%	7
3	Not decided	15%	3
4	Disagree	40%	8
5	Strongly disagree	5%	1

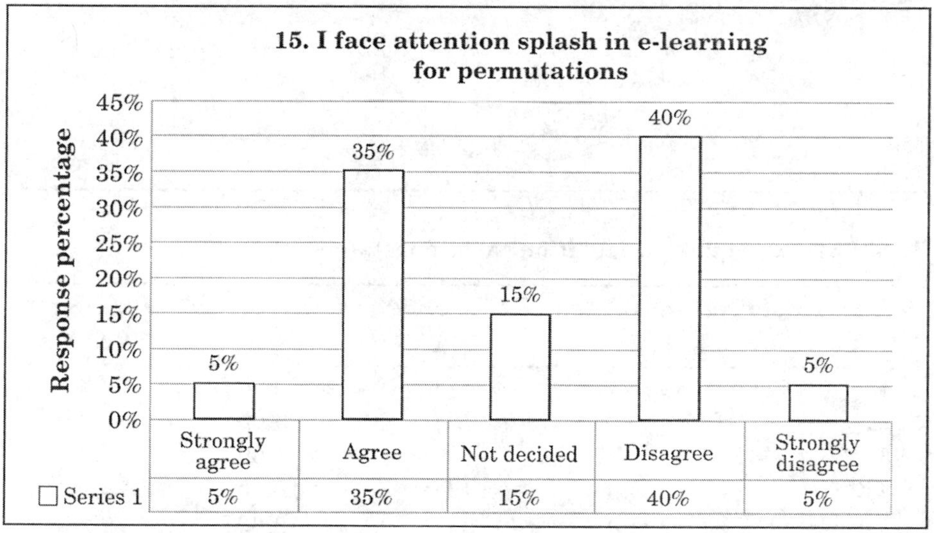

Fig. 15: Graphical representation of statement 15 in percentage.

5% of the students strongly agree with the statement "I face attention splash in e-learning for Permutations".

35% of the students agree with the statement "I face attention splash in e-learning for Permutations".

15% of the students not decided with the statement "I face attention splash in e-learning for Permutations".

40% of the students disagree with the statement "I face attention splash in e-learning for Permutations".

5% of the students strongly disagree with the statement "I face attention splash in e-learning for Permutations".

Data analysis using Chi-square are given in Table 51 and Table 52.

H_0 : Response is uniformly distributed in the 5 point scale

H_1 : Response is not uniformly distributed in the 5 point scale

Table 51: Chi-square table for statement 15.

Points	*Observed frequency*	*Expected frequency*
1	1	4
2	7	4
3	3	4
4	8	4
5	1	4

Table 52: Chi-square calculation table for statement 15.

Points	*Observed*	*Expected*	*Difference*	*Difference sq.*	*Diff. Sq. / Exp fr.*
1	1	4	–3.00	9.00	2.25
2	7	4	3.00	9.00	2.25
3	3	4	–1.00	1.00	0.25
4	8	4	4.00	16.00	4.00
5	1	4	–3.00	9.00	2.25
					11.000

The x^2 value is 11. The P-value is 0.027. The result is significant at $p = \leq 0.05$

Table value of Chi-square at 4 df at 0.05 significance level is 9.49. Calculated value of Chi-square is more than the table value therefore, Null hypothesis is rejected. This revealed that reaction is not uniformly distributed in the 5 point scale. More load is on disagree therefore most of the students disagreed with the statement "I face attention splash in e-learning for Permutations".

Statement 16: I know Education based websites for Permutations.

***Polarity*:** Positive

Responses of the students in percentage is given Table 53 and Fig. 16.

25% of the students strongly agree with the statement "I know Education based websites for Permutations".

65% of the students agree with the statement "I know Education based websites for Permutations".

0% of the students not decided with the statement "I know Education based websites for Permutations".

10% of the students disagree with the statement "I know Education based websites for Permutations".

0.00% of the students strongly disagree with the statement "I know Education based websites for Permutations".

Data analysis using Chi-square are given in Table 54 and Table 55.

H_0 : Response is uniformly distributed in the 5 point scale

H_1 : Response is not uniformly distributed in the 5 point scale

Table 53: Students response in percentage

Points	*Response*	*Response percent*	*Response count*
5	Strongly agree	25%	5
4	Agree	65%	13
3	Not decided	0%	0
2	Disagree	10%	2
1	Strongly disagree	0%	0

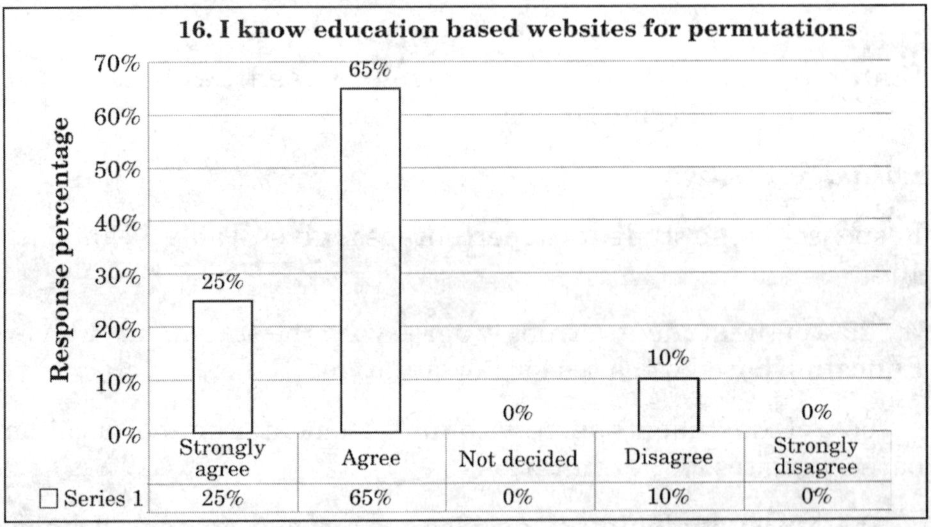

Fig. 16: Graphical representation of statement 16 in percentage.

Table 54: Chi-square table for statement 16.

Points	*Observed frequency*	*Expected frequency*
5	5	4
4	13	4
3	0	4
2	2	4
1	0	4

Table 55: Chi-square calculation table for statement 16.

Points	*Observed*	*Expected*	*Difference*	*Difference sq.*	*Diff. Sq. / Exp fr.*
5	5	4	1.00	1.00	0.25
4	13	4	9.00	81.00	20.25
3	0	4	–4.00	16.00	4.00
2	2	4	–2.00	4.00	1.00
1	0	4	–4.00	16.00	4.00
					29.500

The x^2 value is 29.5. The P-value is <0.001. The result is significant at $p = \leq 0.05$

Table value of Chi-square at 4 df at 0.05 significance level is 9.49. Calculated value of Chi-square is more than the table value therefore, Null hypothesis is rejected. This revealed that reaction is not uniformly distributed in the 5 point scale. More load is on agree therefore most of the students agreed with the statement "I know Education based websites for Permutations."

Statement 17: I face problem in learning Permutations through e-learning material.

***Polarity*:** Negative

Responses of the students in percentage is given in Table 56 and Fig. 17.

Table 56: Students response in percentage.

Points	*Response*	*Response percent*	*Response count*
1	Strongly agree	5%	1
2	Agree	15%	3
3	Not decided	5%	1
4	Disagree	70%	14
5	Strongly disagree	5%	1

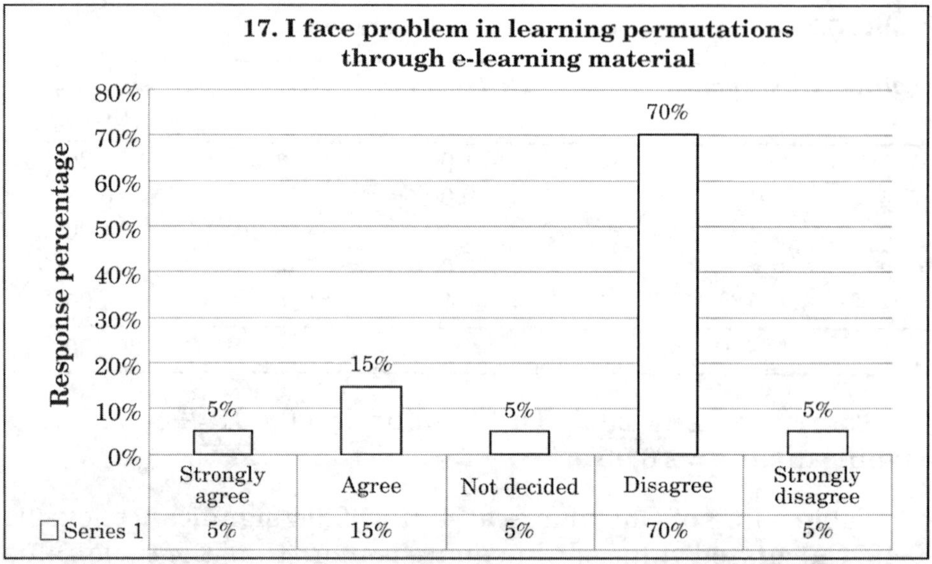

Fig. 17: Graphical representation of statement 17 in percentage.

5% of the students strongly agree with the statement "I face problem in learning Permutations through e-learning material".

15% of the students agree with the statement "I face problem in learning Permutations through e-learning material".

5% of the students not decided with the statement "I face problem in learning Permutations through e-learning material".

70% of the students disagree with the statement "I face problem in learning Permutations through e-learning material".

5% of the students strongly disagree with the statement "I face problem in learning Permutations through e-learning material".

Data analysis using Chi-square are given in Table 57 and Table 58.

H_0 : Response is uniformly distributed in the 5 point scale

H_1 : Response is not uniformly distributed in the 5 point scale

Table 57: Chi-square table for statement 17.

Points	*Observed frequency*	*Expected frequency*
1	1	4
2	3	4
3	1	4
4	14	4
5	1	4

Table 58: Chi-square calculation table for statement 17.

Points	*Observed*	*Expected*	*Difference*	*Difference sq.*	*Diff. Sq. / Exp fr.*
1	1	4	–3.00	9.00	2.25
2	3	4	–1.00	1.00	0.25
3	1	4	–3.00	9.00	2.25
4	14	4	10.00	100.00	25.00
5	1	4	–3.00	9.00	2.25
					32.000

The x^2 value is 32. The P-value is < 0.001. The result is significant at p = ≤ 0.05

Table value of Chi-square at 4 df at 0.05 significance level is 9.49. Calculated value of Chi-square is more than the table value therefore, Null hypothesis is rejected. This revealed that reaction is not uniformly distributed in the 5 point scale. More load is on disagree therefore most of the students agree with the statement "I face problem in learning Permutations through e-learning material".

Statement 18: I get boredom in e-learning on learning Permutations.

***Polarity*:** Negative

Responses of the students in percentage is given in Table 59 and Fig. 18.

Table 59: Students response in percentage

Points	*Response*	*Response percent*	*Response count*
1	Strongly agree	0%	0
2	Agree	35%	7
3	Not decided	25%	5
4	Disagree	35%	7
5	Strongly disagree	5%	1

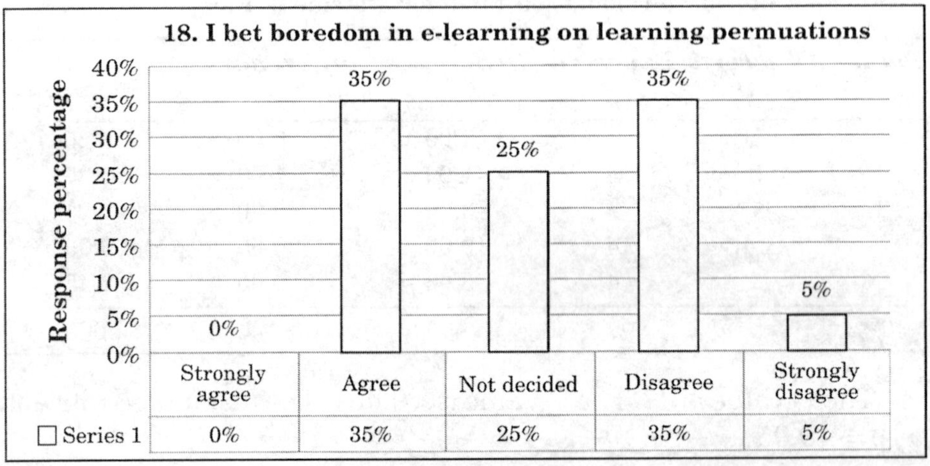

Fig. 18: Graphical representation of statement 18 in percentage.

0% of the students strongly agree with the statement "I get boredom in e-learning on learning Permutations".

35% of the students agree with the statement "I get boredom in e-learning on learning Permutations".

25% of the students not decided with the statement "I get boredom in e-learning on learning Permutations".

35% of the students disagree with the statement "I get boredom in e-learning on learning Permutations".

5% of the students strongly disagree with the statement "I get boredom in e-learning on learning Permutations".

Data analysis using Chi-square are given in Table 60 and Table 61.

H_0: Response is uniformly distributed in the 5 point scale

H_1: Response is not uniformly distributed in the 5 point scale

Table 60: Chi-square table for statement 18.

Points	*Observed frequency*	*Expected frequency*
1	0	4
2	7	4
3	5	4
4	7	4
5	1	4

Table 61: Chi-square calculation table for statement 18.

Points	*Observed*	*Expected*	*Difference*	*Difference sq.*	*Diff. Sq. / Exp fr.*
1	0	4	–4.00	16.00	4.00
2	7	4	3.00	9.00	2.25
3	5	4	1.00	1.00	0.25
4	7	4	3.00	9.00	2.25
5	1	4	–3.00	9.00	2.25
					11.000

The x^2 value is 11. The P-value is 0.027. The result is significant at $p = \leq 0.05$

Table value of Chi-square at 4 df at 0.05 significance level is 9.49. Calculated value of Chi-square is more than the table value therefore, Null hypothesis is rejected. This revealed that reaction is not uniformly distributed in the 5 point students agree as well as disagree with the statement “I get boredom in e-learning on learning Permutations”.

Statement 19: I need teacher's help while learning Permutations through e-learning.

***Polarity*:** Negative

Responses of the students in percentage is given in Table 62 and Fig. 19.

Table 62: Students response in percentage.

Points	*Response*	*Response percent*	*Response count*
1	Strongly agree	20%	4
2	Agree	20%	4
3	Not decided	15%	3
4	Disagree	45%	9
5	Strongly disagree	0%	0

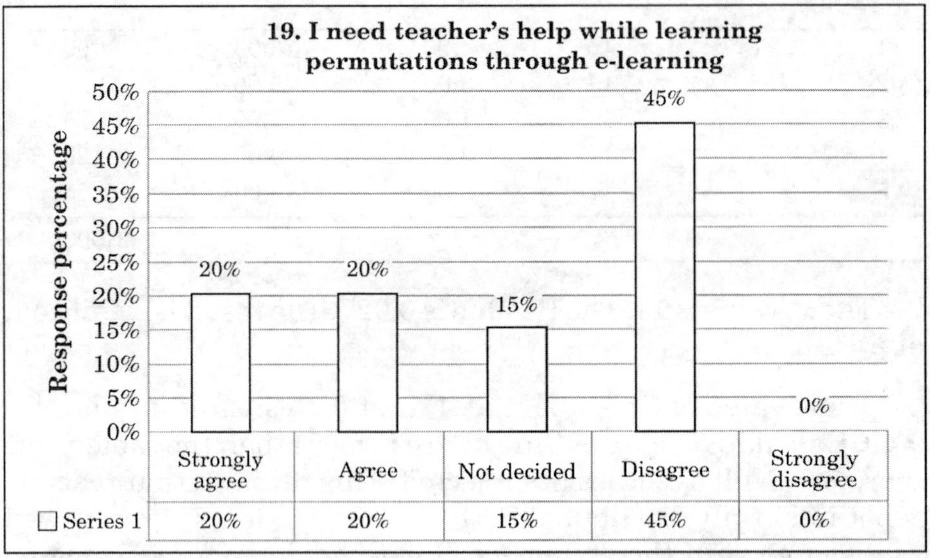

Fig. 19: Graphical representation of statement 19 in percentage.

20% of the students strongly agree with the statement "I need teacher's help while learning Permutations through e-learning".

20% of the students agree with the statement "I need teacher's help while learning Permutations through e-learning".

15% of the students not decided with the statement "I need teacher's help while learning Permutations through e-learning".

45% of the students disagree with the statement "I need teacher's help while learning Permutations through e-learning".

0% of the students strongly disagree with the statement "I need teacher's help while learning Permutations through e-learning".

Data analysis using Chi-square are given in Table 63 and Table 64.

H_0 : Response is uniformly distributed in the 5 point scale

H_1 : Response is not uniformly distributed in the 5 point scale

Table 63: Chi-square table for statement 19.

Points	*Observed frequency*	*Expected frequency*
1	4	4
2	4	4
3	3	4
4	9	4
5	0	4

Table 64: Chi-square calculation table for statement 19.

Points	*Observed*	*Expected*	*Difference*	*Difference sq.*	*Diff. Sq. / Exp fr.*
1	4	4	0.00	0.00	0.00
2	4	4	0.00	0.00	0.00
3	3	4	–1.00	1.00	0.25
4	9	4	5.00	25.00	6.25
5	0	4	–4.00	16.00	4.00
					10.500

The x^2 value is 10.5. The P-value is 0.033. The result is significant at $p = \leq 0.05$

Table value of Chi-square at 4 df at 0.05 significance level is 9.49. Calculated value of Chi-square is more than the table value therefore, Null hypothesis is rejected. This revealed that reaction is not uniformly distributed in the 5 point scale. More load is on disagree therefore most of the students disagreed with the statement "I need teacher's help while learning Permutations through e-learning."

Statement 20: I enjoyed e-learning as I can watch video in the topic Permutations.

***Polarity*:** Positive

Responses of the students in percentage is given in Table 65 and Fig. 20.

30% of the students strongly agree with the statement "I enjoyed e-learning as I can watch video in the topic Permutations".

60% of the students agree with the statement "I enjoyed e-learning as I can watch video in the topic Permutations".

5% of the students not decided with the statement "I enjoyed e-learning as I can watch video in the topic Permutations".

0% of the students disagree with the statement "I enjoyed e-learning as I can watch video in the topic Permutations".

5% of the students strongly disagree with the statement "I enjoyed e-learning as I can watch video in the topic Permutations".

Data analysis using Chi-square are given in Table 66 and Table 67.

H_0 : Response is uniformly distributed in the 5 point scale

H_1 : Response is not uniformly distributed in the 5 point scale

Table 65: Students response in percentage.

Points	*Response*	*Response percent*	*Response count*
5	Strongly agree	30%	6
4	Agree	60%	12
3	Not decided	5%	1
2	Disagree	0%	0
1	Strongly disagree	5%	1

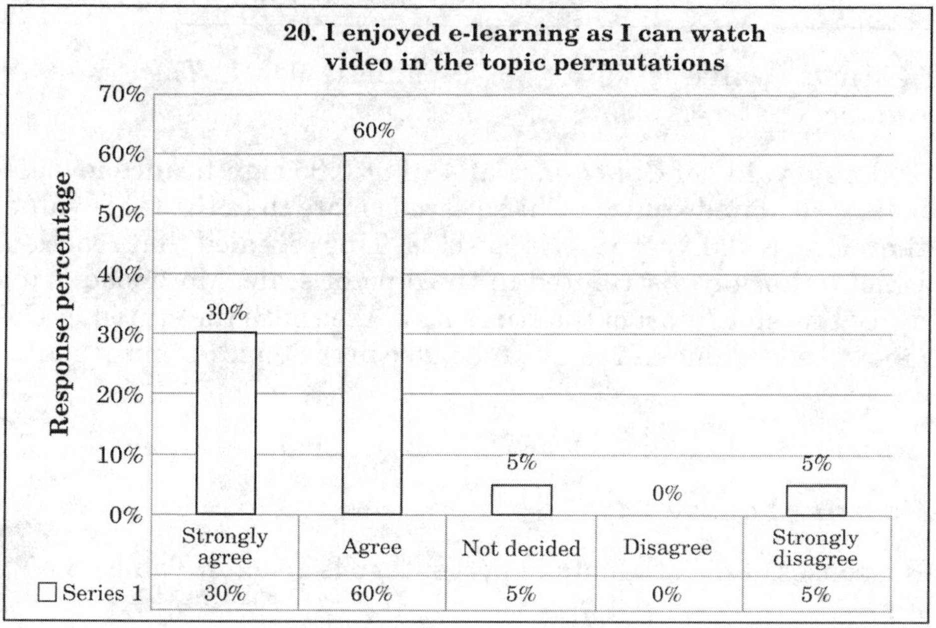

Fig. 20: Graphical representation of statement 20 in percentage.

Table 66: Chi-square table for statement 20.

Points	*Observed frequency*	*Expected frequency*
5	6	4
4	12	4
3	1	4
2	0	4
1	1	4

Table 67: Chi-square calculation table for statement 20.

Points	*Observed*	*Expected*	*Difference*	*Difference sq.*	*Diff. Sq. / Exp fr.*
5	6	4	2.00	4.00	1.00
4	12	4	8.00	64.00	16.00
3	1	4	–3.00	9.00	2.25
2	0	4	–4.00	16.00	4.00
1	1	4	–3.00	9.00	2.25
					25.500

The x^2 value is 25.5. The P-value is 0.001. The result is significant at p = ≤ 0.05.

Table value of Chi-square at 4 df at 0.05 significance level is 9.49. Calculated value of Chi-square is more than the table value therefore, Null hypothesis is rejected. This revealed that reaction is not uniformly distributed in the 5 point scale. More load is on agree therefore most of the students agreed with the statement "I enjoyed e-learning as I can watch video in the topic Permutations."

Statement 21: I like to learn Groups through e-learning.

***Polarity*:** Positive

Responses of the students in percentage is given in Table 68 and Fig. 21.

Table 68: Students response in percentage.

Points	*Response*	*Response percent*	*Response count*
5	Strongly agree	10%	2
4	Agree	65%	13
3	Not decided	10%	2
2	Disagree	15%	3
1	Strongly disagree	0%	0

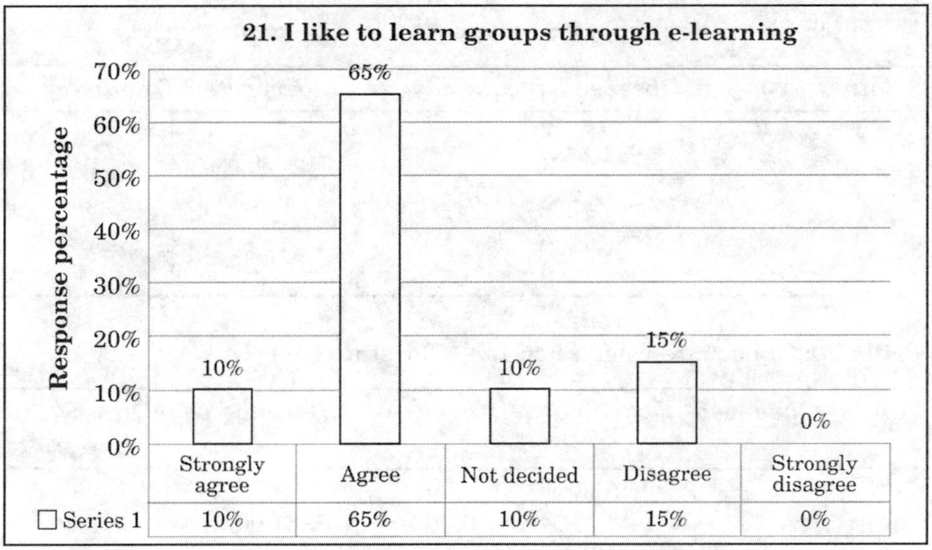

Fig. 21: Graphical representation of statement 21 in percentage.

10% of the students strongly agree with the statement "I like to learn Groups through e-learning".

65% of the students agree with the statement "I like to learn Groups through e-learning."

10% of the students not decided with the statement "I like to learn Groups through e-learning".

15% of the students disagree with the statement "I like to learn Groups through e-learning".

0.00% of the students strongly disagree with the statement "I like to learn Groups through e-learning".

Data analysis using Chi-square are given in Table 69 and Table 70.

H_0: Response is uniformly distributed in the 5 point scale

H_1: Response is not uniformly distributed in the 5 point scale

Table 69: Chi-square table for statement 21.

Points	*Observed frequency*	*Expected frequency*
5	2	4
4	13	4
3	2	4
2	3	4
1	0	4

Table 70: Chi-square calculation table for statement 21.

Points	*Observed*	*Expected*	*Difference*	*Difference sq.*	*Diff. Sq. / Exp fr.*
5	2	4	–2.00	4.00	1.00
4	13	4	9.00	81.00	20.25
3	2	4	–2.00	4.00	1.00
2	3	4	–1.00	1.00	0.25
1	0	4	–4.00	16.00	4.00
					26.500

The x^2 value is 26.5. The P-value is 0.001. The result is significant at $p = \leq 0.05$

Table value of Chi-square at 4 df at 0.05 significance level is 9.49. Calculated value of Chi-square is more than the table value therefore, Null hypothesis is rejected. This revealed that reaction is not uniformly distributed in the 5 point scale. More load is on agree therefore most of the students agree with the statement "I like to learn Groups through e-learning".

Statement 22: I can learn Groups without Teacher and only through e-learning material.

***Polarity*:** Positive

Responses of the students in percentage is given in Table 71 and Fig. 22.

Table 71: Students response in percentage.

Points	*Response*	*Response percent*	*Response count*
5	Strongly agree	10%	2
4	Agree	40%	8
3	Not decided	15%	3
2	Disagree	30%	6
1	Strongly disagree	5%	1

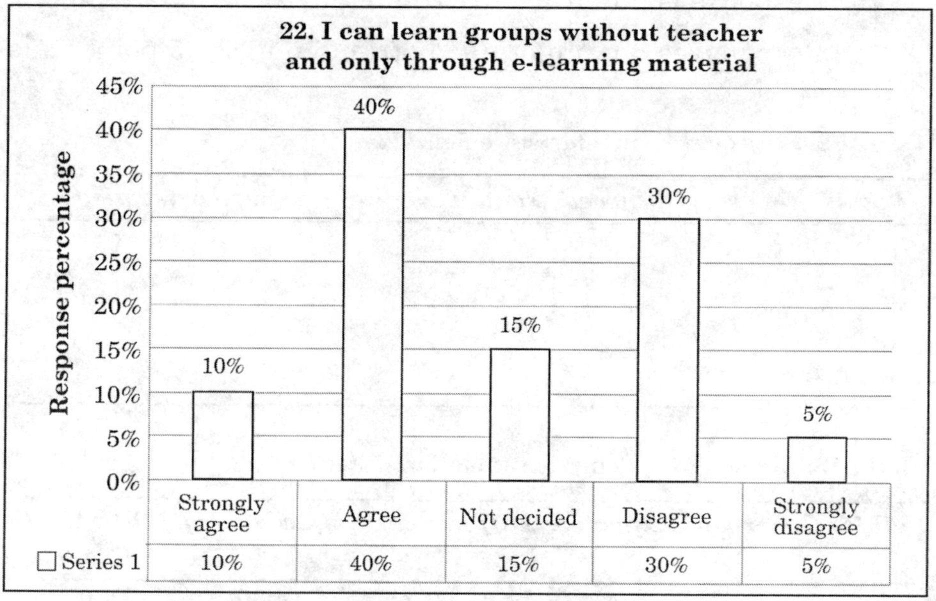

Fig. 22: Graphical representation of statement 22 in percentage.

10% of the students strongly agree with the statement "I can learn Groups without Teacher and only through e-learning material".

40% of the students agree with the statement "I can learn Groups without Teacher and only through e-learning material".

15% of the students not decided with the statement "I can learn Groups without Teacher and only through e-learning material".

30% of the students disagree with the statement "I can learn Groups without Teacher and only through e-learning material".

5% of the students strongly disagree with the statement "I can learn Groups without Teacher and only through e-learning material".

Data analysis using Chi-square are given in Table 72 and Table 73.

H_0 : Response is uniformly distributed in the 5 point scale

H_1 : Response is not uniformly distributed in the 5 point scale

Table 72: Chi-square table for statement 22.

Points	***Observed frequency***	***Expected frequency***
5	2	4
4	8	4
3	3	4
2	6	4
1	1	4

Table 73: Chi-square calculation table for statement 22.

Points	***Observed***	***Expected***	***Difference***	***Difference sq.***	***Diff. Sq. / Exp fr.***
5	2	4	–2.00	4.00	1.00
4	8	4	4.00	16.00	4.00
3	3	4	–1.00	1.00	0.25
2	6	4	2.00	4.00	1.00
1	1	4	–3.00	9.00	2.25
					8.500

The x^2 value is 8.5. The P-value is 0.075. The result is not significant at $p = \leq 0.05$

Table value of Chi-square at 4 df at 0.05 significance level is 9.49. Calculated value of Chi -square is less than the table value therefore, Null hypothesis is not rejected. This revealed that reaction is uniformly distributed in the 5 point scale.

Statement 23: I face problem in computer based learning at Groups.

***Polarity*:** Negative

Responses of the students in percentage is given in Table 74 and Fig. 23.

Table 74: Students response in percentage.

Points	***Response***	***Response percent***	***Response count***
1	Strongly agree	5%	1
2	Agree	30%	6
3	Not decided	25%	5
4	Disagree	40%	8
5	Strongly disagree	0%	0

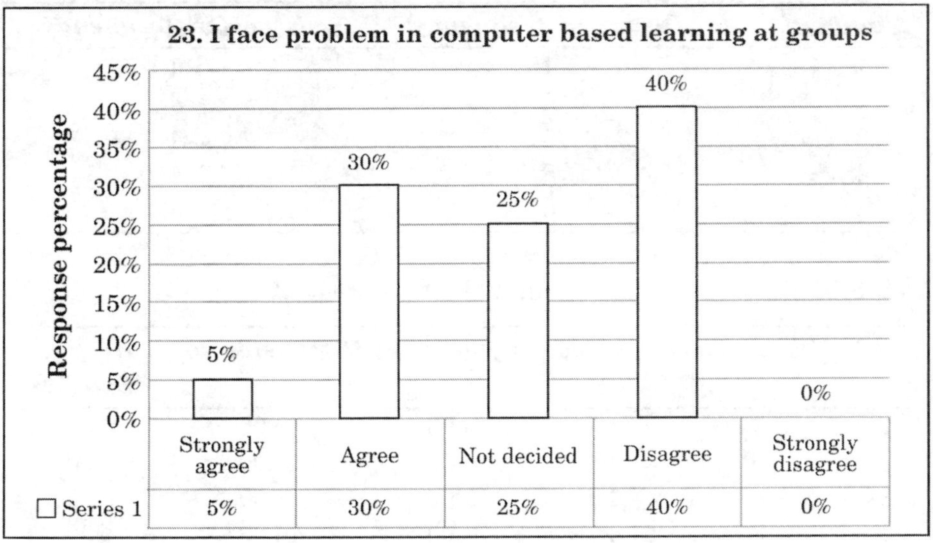

Fig. 23: Graphical representation of statement 23 in percentage.

5% of the students strongly agree with the statement "I face problem in computer based learning at Groups".

30% of the students agree with the statement "I face problem in computer based learning at Groups".

25% of the students not decided with the statement "I face problem in computer based learning at Groups".

40% of the students disagree with the statement "I face problem in computer based learning at Groups".

0.00% of the students strongly disagree with the statement "I face problem in computer based learning at Groups".

Data analysis using Chi-square are given in Table 75 and Table 76.

H_0 : Response is uniformly distributed in the 5 point scale

H_1 : Response is not uniformly distributed in the 5 point scale

Table 75: Chi-square table for statement 23.

Points	*Observed frequency*	*Expected frequency*
1	1	4
2	6	4
3	5	4
4	8	4
5	0	4

Table 76: Chi-square calculation table for statement 23.

Points	*Observed*	*Expected*	*Difference*	*Difference sq.*	*Diff. Sq. / Exp fr.*
1	1	4	–3.00	9.00	2.25
2	6	4	2.00	4.00	1.00
3	5	4	1.00	1.00	0.25
4	8	4	4.00	16.00	4.00
5	0	4	–4.00	16.00	4.00
					11.500

The x^2 value is 11.5. The P-value is 0.021. The result is significant at $p = \leq 0.05$.

Table value of Chi-square at 4 df at 0.05 significance level is 9.49. Calculated value of Chi-square is more than the table value therefore, Null hypothesis is rejected. This revealed that reaction is not uniformly distributed in the 5 point scale. More load is on **disagree** therefore most of the students **disagreed** with the statement "I face problem in computer based learning at Groups".

Statement 24: I face problem in answering Google's Quiz online test in Groups.

***Polarity*:** Negative

Responses of the students in percentage is given in Table 77 and Fig. 24.

Table 77: Students response in percentage.

Points	*Response*	*Response percent*	*Response count*
1	Strongly agree	0%	0
2	Agree	35%	7
3	Not decided	10%	2
4	Disagree	55%	11
5	Strongly disagree	0%	0

0% of the students strongly agree with the statement "I face problem in answering Google's Quiz online test in Groups".

35% of the students agree with the statement "I face problem in answering Google's Quiz online test in Groups".

10% of the students not decided with the statement "I face problem in answering Google's Quiz online test in Groups".

55% of the students disagree with the statement "I face problem in answering Google's Quiz online test in Groups".

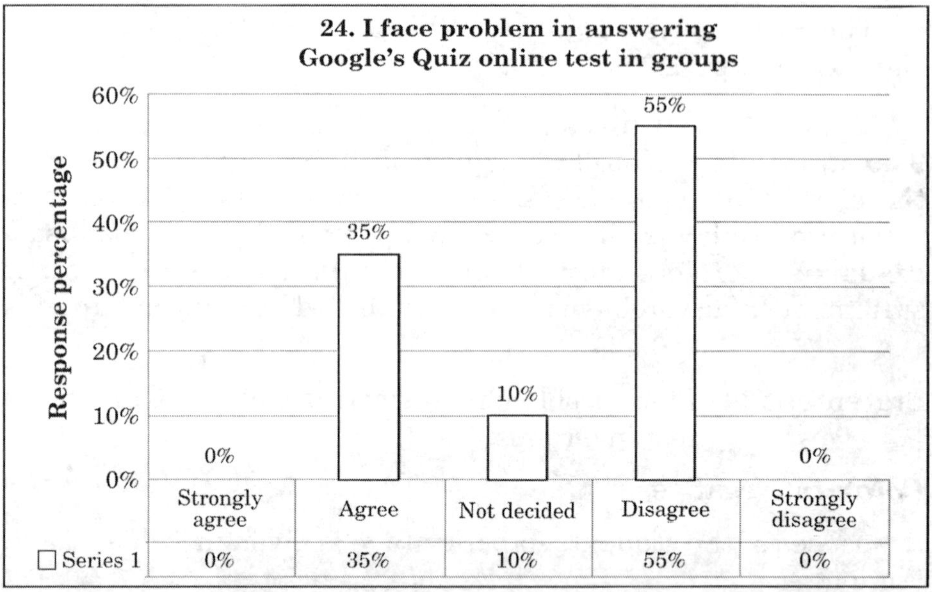

Fig. 24: Graphical representation of statement 24 in percentage.

0.00% of the students strongly disagree with the statement "I face problem in answering Google's Quiz online test in Groups".

Data analysis using Chi-square are given in Table 78 and Table 79.

H_0 : Response is uniformly distributed in the 5 point scale

H_1 : Response is not uniformly distributed in the 5 point scale

Table 78: Chi-square table for statement 24.

Points	*Observed frequency*	*Expected frequency*
1	0	4
2	7	4
3	2	4
4	11	4
5	0	4

Table 79: Chi-square calculation table for statement 24.

Points	***Observed***	***Expected***	***Difference***	***Difference sq.***	***Diff. Sq. / Exp fr.***
1	0	4	–4.00	16.00	4.00
2	7	4	3.00	9.00	2.25
3	2	4	–2.00	4.00	1.00
4	11	4	7.00	49.00	12.25
5	0	4	–4.00	16.00	4.00
					23.500

The x^2 value is 23.5. The P-value is < 0.001. The result is significant at p = ≤ 0.05

Table value of Chi-square at 4 df at 0.05 significance level is 9.49. Calculated value of Chi-square is more than the table value therefore, Null hypothesis is rejected. This revealed that reaction is not uniformly distributed in the 5-point scale. More load is on **disagree** therefore most of the students **disagreed** with the statement "I face problem in answering Google's Quiz online test in Groups".

Statement 25: I face attention splash in e-learning for Groups.

***Polarity*:** Negative

Responses of the students in percentage is given in Table 80 and Fig. 25.

Table 80: Students response in percentage.

Points	***Response***	***Response percent***	***Response count***
1	Strongly agree	5%	1
2	Agree	35%	7
3	Not decided	20%	4
4	Disagree	40%	8
5	Strongly disagree	0%	0

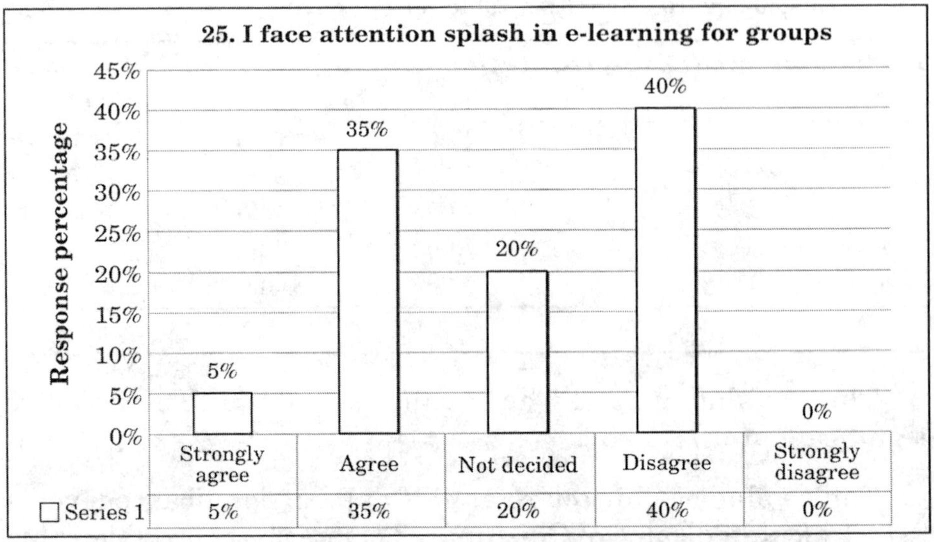

Fig. 25: Graphical representation of statement 25 in percentage.

5% of the students strongly agree with the statement "I face attention splash in e-learning for Groups".

35% of the students agree with the statement "I face attention splash in e-learning for Groups".

20% of the students not decided with the statement "I face attention splash in e-learning for Groups".

40% of the students disagree with the statement "I face attention splash in e-learning for Groups".

0.00% of the students strongly disagree with the statement "I face attention splash in e-learning for Groups".

Data analysis using Chi-square are given in Table 81 and Table 82.

H_0 : Response is uniformly distributed in the 5 point scale

H_1 : Response is not uniformly distributed in the 5 point scale

Table 81: Chi-square table for statement 25.

Points	*Observed frequency*	*Expected frequency*
1	1	4
2	7	4
3	4	4
4	8	4
5	0	4

Table 82: Chi-square calculation table for statement 25.

Points	*Observed*	*Expected*	*Difference*	*Difference sq.*	*Diff. Sq. / Exp fr.*
1	1	4	–3.00	9.00	2.25
2	7	4	3.00	9.00	2.25
3	4	4	0.00	0.00	0.00
4	8	4	4.00	16.00	4.00
5	0	4	–4.00	16.00	4.00
					12.500

The x^2 value is 12.5. The P-value is 0.014. The result is significant at p = ≤ 0.05

Table value of Chi-square at 4 df at 0.05 significance level is 9.49. Calculated value of Chi-square is more than the table value therefore, Null hypothesis is rejected. This revealed that reaction is not uniformly distributed in the 5 point scale. More load is on **disagree** therefore most of the students **disagreed** with the statement “I face attention splash in e-learning for Groups”.

Statement 26: I know Education based websites for Groups.

***Polarity*:** Positive

Responses of the students in percentage is given in Table 83 and Fig. 26.

Table 83: Students response in percentage.

Points	*Response*	*Response percent*	*Response count*
5	Strongly agree	10%	2
4	Agree	65%	13
3	Not decided	20%	4
2	Disagree	5%	1
1	Strongly disagree	0%	0

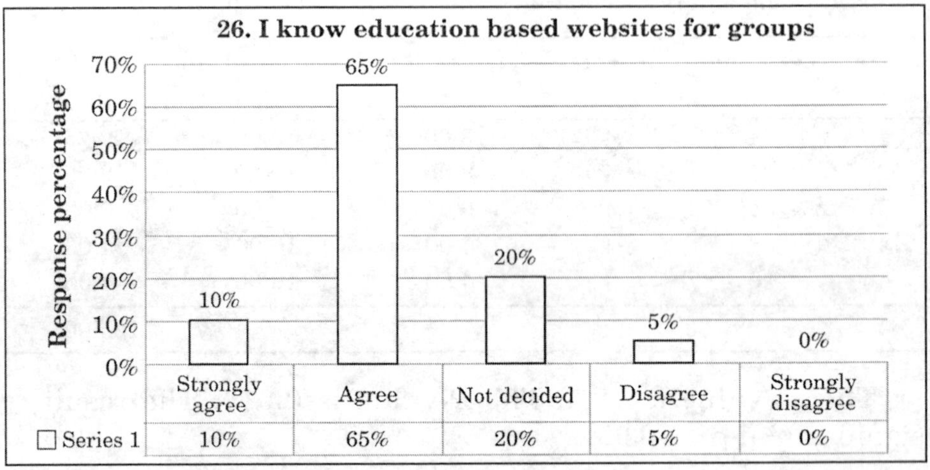

Fig. 26: Graphical representation of statement 26 in percentage.

10% of the students strongly agree with the statement "I know Education based websites for Groups".

65% of the students agree with the statement "I know Education based websites for Groups".

20% of the students not decided with the statement "I know Education based websites for Groups".

5% of the students disagree with the statement "I know Education based websites for Groups".

0.00% of the students strongly disagree with the statement "I know Education based websites for Groups".

Data analysis using Chi-square are given in Table 84 and Table 85.

H_0: Response is uniformly distributed in the 5 point scale

H_1: Response is not uniformly distributed in the 5 point scale

Table 84: Chi-square table for statement 26.

Points	*Observed frequency*	*Expected frequency*
5	2	4
4	13	4
3	4	4
2	1	4
1	0	4

Table 85: Chi-square calculation table for statement 26.

Points	*Observed*	*Expected*	*Difference*	*Difference sq.*	*Diff. Sq. / Exp fr.*
5	2	4	–2.00	4.00	1.00
4	13	4	9.00	81.00	20.25
3	4	4	0.00	0.00	0.00
2	1	4	–3.00	9.00	2.25
1	0	4	–4.00	16.00	4.00
					27.500

The x^2 value is 27.5. The P-value is 0.001. The result is significant at $p = \leq 0.05$

Table value of Chi-square at 4 df at 0.05 significance level is 9.49. Calculated value of Chi-square is more than the table value therefore, Null hypothesis is rejected. This revealed that reaction is not uniformly distributed in the 5 point scale. More load is on **agree** therefore most of the students **agreed** with the statement "I know Education based websites for Groups".

Statement 27: I face problem in learning Groups through e-learning material.

***Polarity*:** Negative

Responses of the students in percentage is given in Table 86 and Fig. 27.

Table 86: Students response in percentage.

Points	*Response*	*Response percent*	*Response count*
1	Strongly agree	5%	1
2	Agree	40%	8
3	Not decided	10%	2
4	Disagree	45%	9
5	Strongly disagree	0%	0

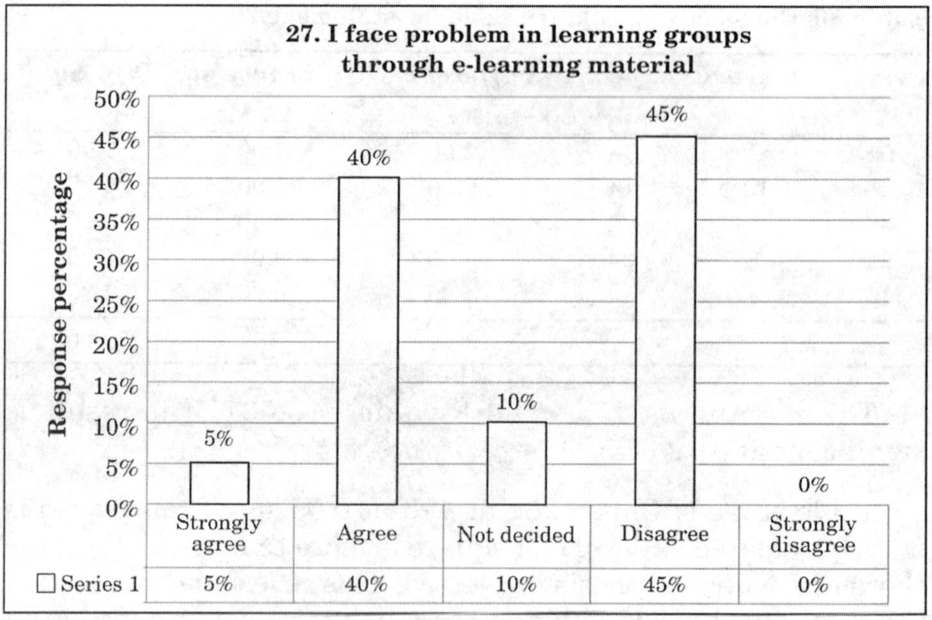

Fig. 27: Graphical representation of statement 27 in percentage.

5% of the students strongly agree with the statement "I face problem in learning Groups through e-learning material".

40% of the students agree with the statement "I face problem in learning Groups through e-learning material".

10% of the students not decided with the statement "I face problem in learning Groups through e-learning material".

45% of the students disagree with the statement "I face problem in learning Groups through e-learning material".

0.00% of the students strongly disagree with the statement "I face problem in learning Groups through e-learning material".

Data analysis using Chi-square are given in Table 87 and Table 88.

H_0 : Response is uniformly distributed in the 5 point scale

H_1 : Response is not uniformly distributed in the 5 point scale

Table 87: Chi-square table for statement 27.

Points	*Observed frequency*	*Expected frequency*
1	1	4
2	8	4
3	2	4
4	9	4
5	0	4

Table 88: Chi-square calculation table for statement 27.

Points	*Observed*	*Expected*	*Difference*	*Difference sq.*	*Diff. Sq. / Exp fr.*
1	1	4	–3.00	9.00	2.25
2	8	4	4.00	16.00	4.00
3	2	4	–2.00	4.00	1.00
4	9	4	5.00	25.00	6.25
5	0	4	–4.00	16.00	4.00
					17.500

The x^2 value is 17.5. The P-value is 0.002. The result is significant at $p = \leq 0.05$

Table value of Chi-square at 4 df at 0.05 significance level is 9.49. Calculated value of Chi-square is more than the table value therefore, Null hypothesis is rejected. This revealed that reaction is not uniformly distributed in the 5 point scale. More load is on **disagree** therefore most of the students **disagreed** with the statement "I face problem in learning Groups through e-learning material."

Statement 28: I get boredom in e-learning on learning Groups.

***Polarity*:** Negative

Responses of the students in percentage is given in Table 89 and Fig. 28.

Table 89: Students response in percentage.

Points	*Response*	*Response percent*	*Response count*
1	Strongly agree	0%	0
2	Agree	45%	9
3	Not decided	10%	2
4	Disagree	45%	9
5	Strongly disagree	0%	0

0% of the students strongly agree with the statement "I get boredom in e-learning on learning Groups".

45% of the students agree with the statement "I get boredom in e-learning on learning Groups".

10% of the students not decided with the statement "I get boredom in e-learning on learning Groups".

45% of the students disagree with the statement "I get boredom in e-learning on learning Groups".

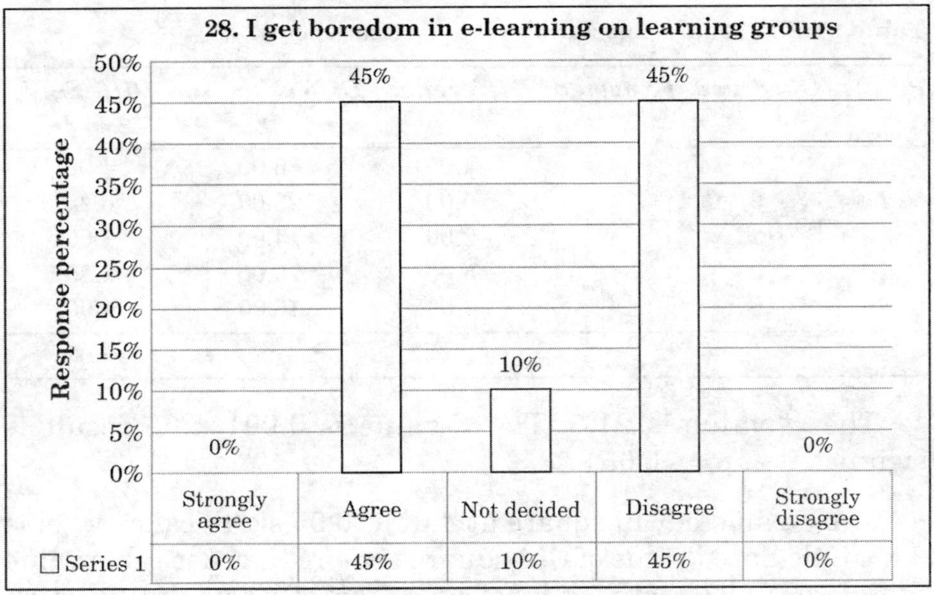

Fig. 28: Graphical representation of statement 28 in percentage.

0.00% of the students strongly disagree with the statement "I get boredom in e-learning on learning Groups".

Data analysis using Chi-square are given in Table 90 and Table 91.

H_0 : Response is uniformly distributed in the 5 point scale

H_1 : Response is not uniformly distributed in the 5 point scale

Table 90: Chi-square table for statement 28

Points	*Observed frequency*	*Expected frequency*
1	0	4
2	9	4
3	2	4
4	9	4
5	0	4

Table 91: Chi-square calculation table for statement 28.

Points	*Observed*	*Expected*	*Difference*	*Difference sq.*	*Diff. Sq. / Exp fr.*
1	0	4	–4.00	16.00	4.00
2	9	4	5.00	25.00	6.25
3	2	4	–2.00	4.00	1.00
4	9	4	5.00	25.00	6.25
5	0	4	–4.00	16.00	4.00
					21.500

The x^2 value is 21.5. The P-value is 0.001. The result is significant at $p = \leq 0.05$

Table value of Chi-square at 4 df at 0.05 significance level is 9.49. Calculated value of Chi-square is more than the table value therefore, Null hypothesis is rejected. This revealed that reaction is not uniformly distributed in the 5 point scale. Equal load is on **agree** and **disagree** therefore equal number of the students **agreed** as well as **disagreed** with the statement "I get boredom in e-learning on learning Groups".

Statement 29: I need teacher's help while learning Groups through e-learning.

***Polarity*:** Negative

Responses of the students in percentage is given Table 92 and Fig. 29.

Table 92: Students response in percentage.

Points	*Response*	*Response percent*	*Response count*
1	Strongly agree	35%	7
2	Agree	45%	9
3	Not decided	0%	0
4	Disagree	20%	4
5	Strongly disagree	0%	0

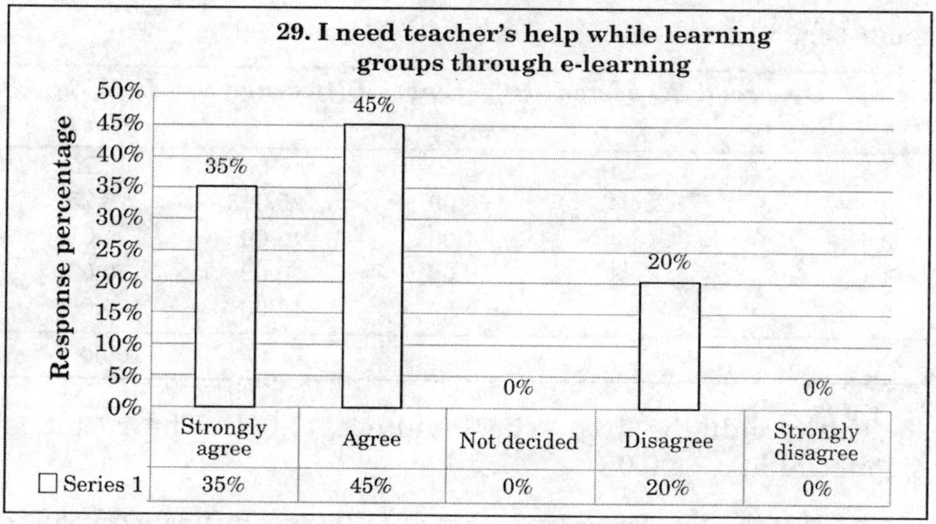

Fig. 29: Graphical representation of statement 29 in percentage.

Data analysis using Chi-square are given in Table 93 and Table 94.

H_0: Response is uniformly distributed in the 5 point scale

H_1: Response is not uniformly distributed in the 5 point scale

Table 93: Chi-square table for statement 29.

Points	*Observed frequency*	*Expected frequency*
1	7	4
2	9	4
3	0	4
4	4	4
5	0	4

Table 94: Chi-square calculation table for statement 29.

Points	*Observed*	*Expected*	*Difference*	*Difference sq.*	*Diff. Sq. / Exp fr.*
1	7	4	3.00	9.00	2.25
2	9	4	5.00	25.00	6.25
3	0	4	–4.00	16.00	4.00
4	4	4	0.00	0.00	0.00
5	0	4	–4.00	16.00	4.00
					16.500

The x^2 value is 16.5. The P-value is 0.002. The result is significant at $p = \leq 0.05$

Table value of Chi-square at 4 df at 0.05 significance level is 9.49. Calculated value of Chi-square is more than the table value therefore, Null hypothesis is rejected. This revealed that reaction is not uniformly distributed in the 5 point scale. More load is on agree therefore most of the students agree with the statement "I need teacher's help while learning Groups through e-learning".

Statement 30: I enjoyed e-learning as I can watch video in the topic Groups.

***Polarity*:** Positive

Responses of the students in percentage is given in Table 95 and Fig. 30.

Table 95: Students response in percentage.

Points	*Response*	*Response percent*	*Response count*
5	Strongly agree	25%	5
4	Agree	50%	10
3	Not decided	10%	2
2	Disagree	10%	2
1	Strongly disagree	5%	1

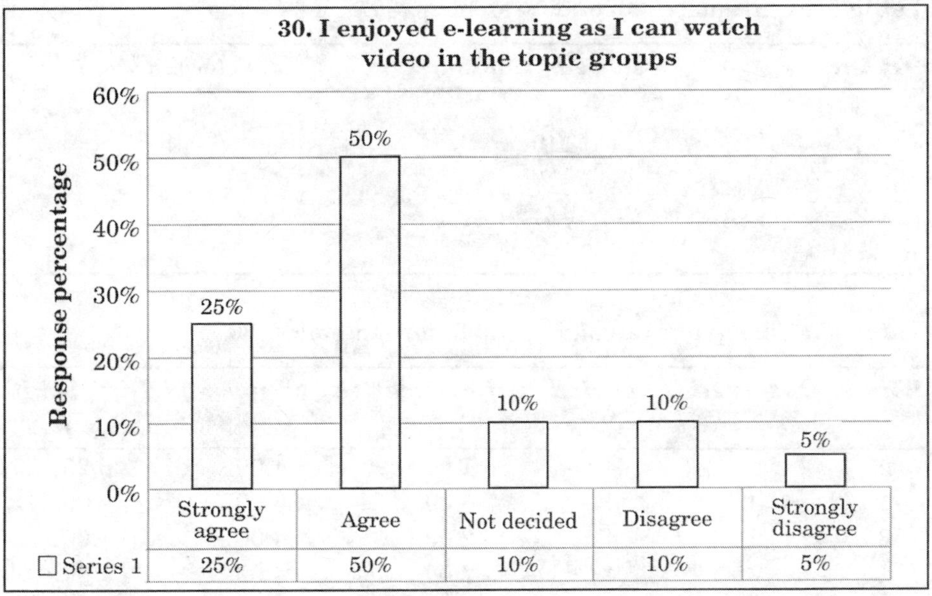

Fig. 30: Graphical representation of statement 30 in percentage.

25% of the students strongly agree with the statement "I enjoyed e-learning as I can watch video in the topic Groups".

50% of the students agree with the statement "I enjoyed e-learning as I can watch video in the topic Groups".

10% of the students not decided with the statement "I enjoyed e-learning as I can watch video in the topic Groups".

10% of the students disagree with the statement "I enjoyed e-learning as I can watch video in the topic Groups".

5% of the students strongly disagree with the statement "I enjoyed e-learning as I can watch video in the topic Groups".

Data analysis using Chi-square are given in Table 96 and Table 97.

H_0: Response is uniformly distributed in the 5 point scale

H_1: Response is not uniformly distributed in the 5 point scale

Table 96: Chi-square table for statement 30.

Points	*Observed frequency*	*Expected frequency*
5	5	4
4	10	4
3	2	4
2	2	4
1	1	4

Table 97: Chi-square calculation table for statement 30.

Points	*Observed*	*Expected*	*Difference*	*Difference sq.*	*Diff. Sq. / Exp fr.*
5	5	4	1.00	1.00	0.25
4	10	4	6.00	36.00	9.00
3	2	4	–2.00	4.00	1.00
2	2	4	–2.00	4.00	1.00
1	1	4	–3.00	9.00	2.25
					13.500

The x^2 value is 13.5. The P-value is 0.009. The result is significant at $p = \leq 0.05$

Table value of Chi-square at 4 df at .05 significance level is 9.49. Calculated value of Chi-square is more than the table value therefore, Null hypothesis is rejected. This revealed that reaction is not uniformly distributed in the 5 point scale. More load is on **agree** therefore most of the students **agreed** with the statement "I enjoyed e-learning as I can watch video in the topic Groups."

Statement 31: I like to learn Equivalence Relations through e-learning.

***Polarity*:** Positive

Responses of the students in percentage is given in Table 98 and Fig. 31.

Table 98: Students response in percentage.

Points	***Response***	***Response percent***	***Response count***
5	Strongly agree	10%	2
4	Agree	75%	15
3	Not decided	10%	2
2	Disagree	5%	1
1	Strongly disagree	0%	0

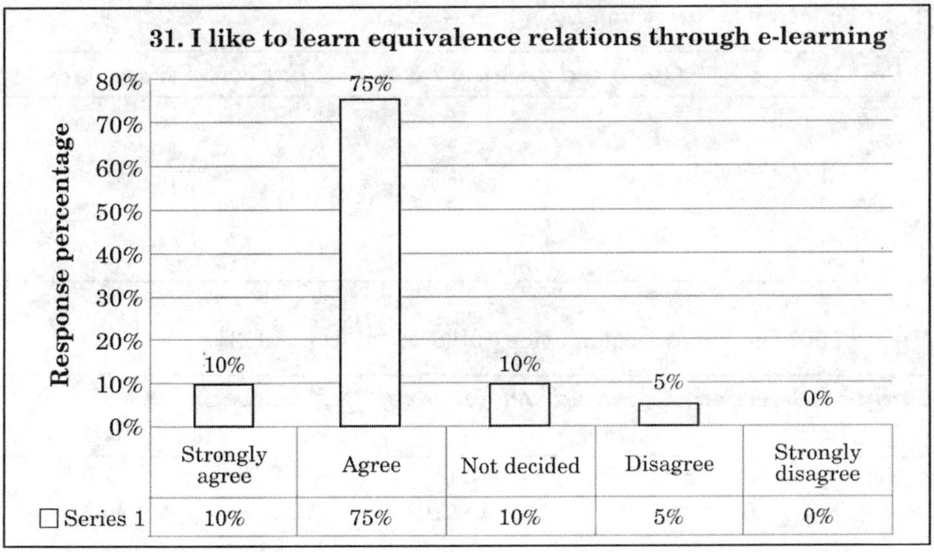

Fig. 31: Graphical representation of statement 31 in percentage.

10% of the students strongly agree with the statement “I like to learn Equivalence Relations through e-learning”.

75% of the students agree with the statement “I like to learn Equivalence Relations through e-learning”.

10% of the students not decided with the statement “I like to learn Equivalence Relations through e-learning”.

5% of the students disagree with the statement “I like to learn Equivalence Relations through e-learning”.

0.00% of the students strongly disagree with the statement "I like to learn Equivalence Relations through e-learning".

Data analysis using Chi-square are given in Table 99 and Table 100.

H_0 : Response is uniformly distributed in the 5 point scale

H_1 : Response is not uniformly distributed in the 5 point scale

Table 99: Chi-square table for statement 31.

Points	*Observed frequency*	*Expected frequency*
5	2	4
4	15	4
3	2	4
2	1	4
1	0	4

Table 100: Chi-square calculation table for statement 31.

Points	*Observed*	*Expected*	*Difference*	*Difference sq.*	*Diff. Sq. / Exp fr.*
5	2	4	–2.00	4.00	1.00
4	15	4	11.00	121.00	30.25
3	2	4	–2.00	4.00	1.00
2	1	4	–3.00	9.00	2.25
1	0	4	–4.00	16.00	4.00
					38.500

The x^2 value is 38.5. The P-value is 0.001. The result is significant at $p = \leq 0.05$

Table value of Chi-square at 4 df at 0.05 significance level is 9.49. Calculated value of Chi-square is more than the table value therefore, Null hypothesis is rejected. This revealed that reaction is not uniformly distributed in the 5 point scale. More load is on **agree** therefore most of the students **agreed** with the statement "I like to learn Equivalence Relations through e-learning".

Statement 32: I can learn Equivalence Relations without Teacher and only through e-learning material.

***Polarity*:** Positive

Responses of the students in percentage is given in Table 101 and Fig. 32.

Table 101: Students response in percentage.

Points	*Response*	*Response percent*	*Response count*
5	Strongly agree	5%	1
4	Agree	55%	11
3	Not decided	10%	2
2	Disagree	20%	4
1	Strongly disagree	10%	2

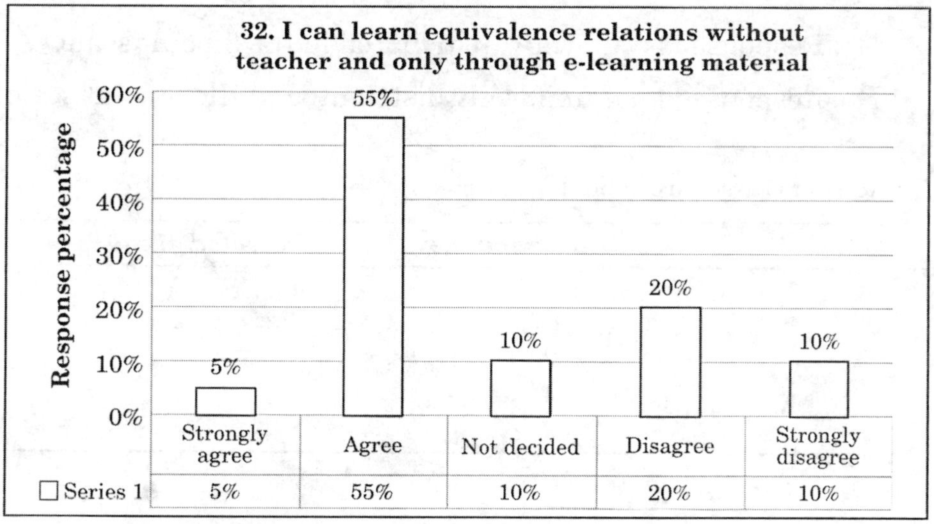

Fig. 32: Graphical representation of statement 32 in percentage.

5% of the students strongly agree with the statement "I can learn Equivalence Relations without Teacher and only through e-learning material".

55% of the students agree with the statement “I can learn Equivalence Relations without Teacher and only through e-learning material”.

10% of the students not decided with the statement “I can learn Equivalence Relations without Teacher and only through e-learning material”.

20% of the students disagree with the statement “I can learn Equivalence Relations without Teacher and only through e-learning material”.

10% of the students strongly disagree with the statement “I can learn Equivalence Relations without Teacher and only through e-learning material”.

Data analysis using Chi-square are given in Table 102 and Table 103.

H_0 : Response is uniformly distributed in the 5 point scale

H_1 : Response is not uniformly distributed in the 5 point scale

Table 102: Chi-square table for statement 32.

Points	*Observed frequency*	*Expected frequency*
5	1	4
4	11	4
3	2	4
2	4	4
1	2	4

Table 103: Chi-square calculation table for statement 32.

Points	*Observed*	*Expected*	*Difference*	*Difference sq.*	*Diff. Sq. / Exp fr.*
5	1	4	–3.00	9.00	2.25
4	11	4	7.00	49.00	12.25
3	2	4	–2.00	4.00	1.00
2	4	4	0.00	0.00	0.00
1	2	4	–2.00	4.00	1.00
					16.500

The x^2 value is 16.5. The P-value is 0.002. The result is significant at $p = \leq 0.05$

Table value of Chi-square at 4 df at 0.05 significance level is 9.49. Calculated value of Chi-square is more than the table value therefore, Null hypothesis is rejected. This revealed that reaction is not uniformly distributed in the 5 point scale. More load is on **agree** therefore most of the students **agreed** with the statement "I can learn Equivalence Relations without Teacher and only through e-learning material".

Statement 33: I face problem in computer based learning at Equivalence Relations.

***Polarity*:** Negative

Responses of the students in percentage is given in Table 104 and Fig. 33.

Table 104: Students response in percentage.

Points	*Response*	*Response percent*	*Response count*
1	Strongly agree	5%	1
2	Agree	25%	5
3	Not decided	10%	2
4	Disagree	60%	12
5	Strongly disagree	0%	0

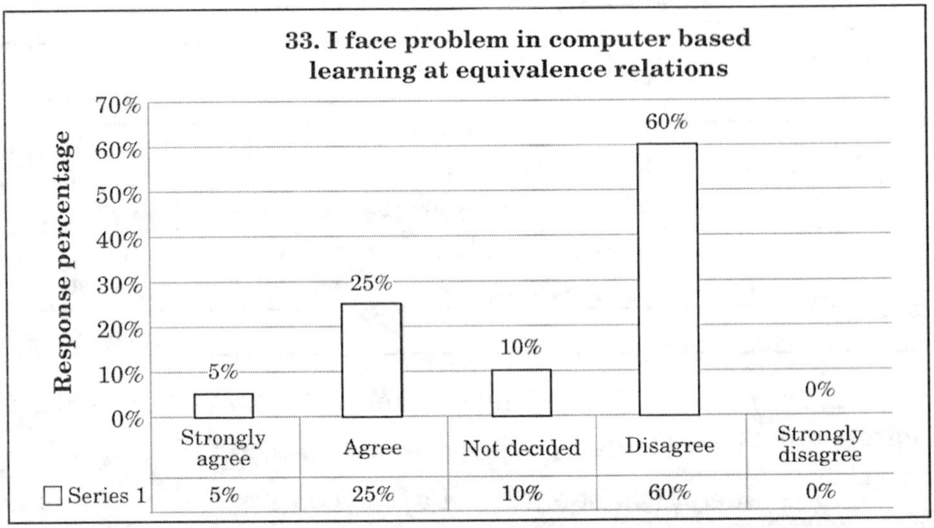

Fig. 33: Graphical representation of statement 33 in percentage.

5% of the students strongly agree with the statement. "I face problem in computer based learning at Equivalence Relations".

25% of the students agree with the statement. "I face problem in computer based learning at Equivalence Relations".

10% of the students not decided with the statement. "I face problem in computer based learning at Equivalence Relations".

60% of the students disagree with the statement. "I face problem in computer based learning at Equivalence Relations".

0.00% of the students strongly disagree with the statement. "I face problem in computer based learning at Equivalence Relations".

Data analysis using Chi-square are given in Table 105 and Table 106.

H_0 : Response is uniformly distributed in the 5 point scale

H_1 : Response is not uniformly distributed in the 5 point scale

Table 105: Chi-square table for statement 33.

Points	*Observed frequency*	*Expected frequency*
1	1	4
2	5	4
3	2	4
4	12	4
5	0	4

Table 106: Chi-square calculation table for statement 33.

Points	*Observed*	*Expected*	*Difference*	*Difference sq.*	*Diff. Sq. / Exp fr.*
1	1	4	–3.00	9.00	2.25
2	5	4	1.00	1.00	0.25
3	2	4	–2.00	4.00	1.00
4	12	4	8.00	64.00	16.00
5	0	4	–4.00	16.00	4.00
					23.500

The x^2 value is 23.5. The P-value is 0.001. The result is significant at $p = \leq 0.05$

Table value of Chi-square at 4 df at 0.05 significance level is 9.49. Calculated value of Chi-square is more than the table value therefore, Null hypothesis is rejected. This revealed that reaction is not uniformly distributed in the 5 point scale. More load is on **disagree** therefore most of the students **disagreed** with the statement "I face problem in computer based learning at Equivalence Relations".

Statement 34: I face problem in answering Google's Quiz online test in Equivalence Relations.

***Polarity*:** Negative

Responses of the students in percentage is given in Table 107 and Fig. 34.

Table 107: Students response in percentage.

Points	*Response*	*Response percent*	*Response count*
1	Strongly agree	0%	0
2	Agree	20%	4
3	Not decided	15%	3
4	Disagree	65%	13
5	Strongly disagree	0%	0

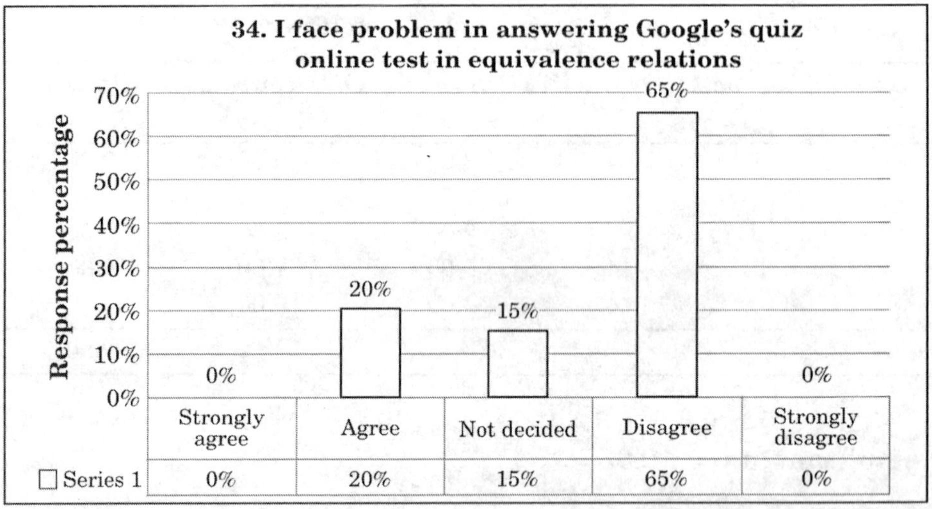

Fig. 34: Graphical representation of statement 34 in percentage.

0% of the students strongly agree with the statement "I face problem in answering Google's Quiz online test in Equivalence Relations".

20% of the students agree with the statement "I face problem in answering Google's Quiz online test in Equivalence Relations".

15% of the students not decided with the statement "I face problem in answering Google's Quiz online test in Equivalence Relations".

65% of the students disagree with the statement "I face problem in answering Google's Quiz online test in Equivalence Relations".

0.00% of the students strongly disagree with the statement "I face problem in answering Google's Quiz online test in Equivalence Relations".

Data analysis using Chi-square are given in Table 108 and Table 109.

H_0 : Response is uniformly distributed in the 5 point scale

H_1 : Response is not uniformly distributed in the 5 point scale

Table 108: Chi-square table for statement 34.

Points	*Observed frequency*	*Expected frequency*
1	0	4
2	4	4
3	3	4
4	13	4
5	0	4

Table 109: Chi-square calculation table for statement 34.

Points	*Observed*	*Expected*	*Difference*	*Difference sq.*	*Diff. Sq. / Exp fr.*
1	0	4	–4.00	16.00	4.00
2	4	4	0.00	0.00	0.00
3	3	4	–1.00	1.00	0.25
4	13	4	9.00	81.00	20.25
5	0	4	–4.00	16.00	4.00
					28.500

The x^2 value is 28.5. The P-value is <0.001. The result is significant at $p = \leq 0.05$

Table value of Chi-square at 4 df at 0.05 significance level is 9.49. Calculated value of Chi-square is more than the table value therefore, Null hypothesis is rejected. This revealed that reaction is not uniformly distributed in the 5 point scale. More load is on **disagree** therefore most of the students **disagreed** with the statement "I face problem in answering Google's Quiz online test in Equivalence Relations".

Statement 35: I face attention splash in e-learning for Equivalence Relations.

***Polarity*:** Negative

Responses of the students in percentage is given in Table 110 and Fig. 35.

Table 110: Students response in percentage.

Points	*Response*	*Response percent*	*Response count*
1	Strongly agree	5%	1
2	Agree	40%	8
3	Not decided	10%	2
4	Disagree	45%	9
5	Strongly disagree	0%	0

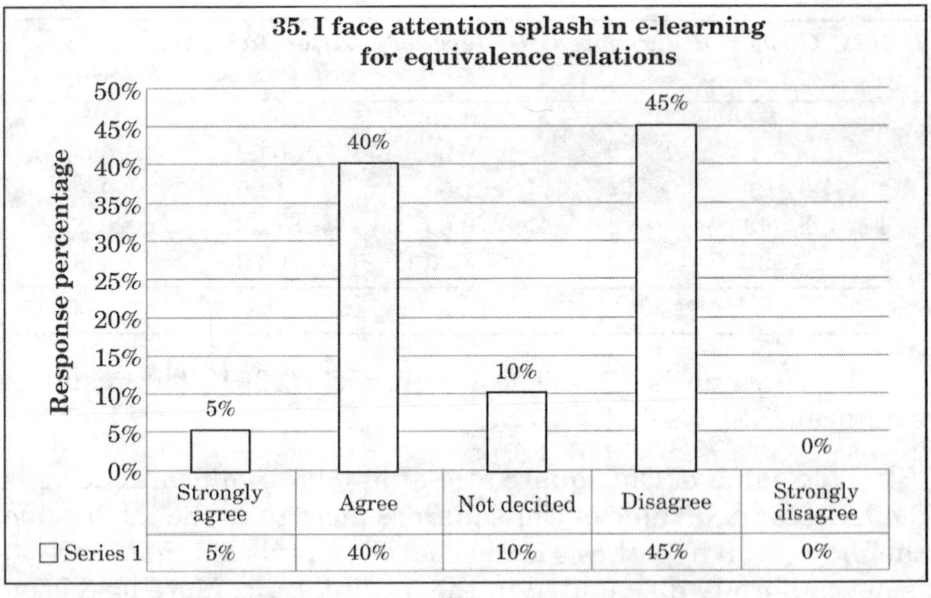

Fig. 35: Graphical representation of statement 35 in percentage.

5% of the students strongly agree with the statement “I face attention splash in e-learning for Equivalence Relations”.

40% of the students agree with the statement “I face attention splash in e-learning for Equivalence Relations”.

10% of the students not decided with the statement “I face attention splash in e-learning for Equivalence Relations”.

45% of the students disagree with the statement “I face attention splash in e-learning for Equivalence Relations”.

0.00% of the students strongly disagree with the statement “I face attention splash in e-learning for Equivalence Relations”.

Data analysis using Chi-square are given in Table 111 and Table 112.

H_0 : Response is uniformly distributed in the 5 point scale

H_1 : Response is not uniformly distributed in the 5 point scale

Table 111: Chi-square table for statement 35.

Points	*Observed frequency*	*Expected frequency*
1	1	4
2	8	4
3	2	4
4	9	4
5	0	4

Table 112: Chi-square calculation table for statement 35.

Points	*Observed*	*Expected*	*Difference*	*Difference sq.*	*Diff. Sq. / Exp fr.*
1	1	4	–3.00	9.00	2.25
2	8	4	4.00	16.00	4.00
3	2	4	–2.00	4.00	1.00
4	9	4	5.00	25.00	6.25
5	0	4	–4.00	16.00	4.00
					17.500

The x^2 value is 17.5. The P-value is 0.002. The result is significant at $p = \leq 0.05$

Table value of Chi-square at 4 df at 0.05 significance level is 9.49. Calculated value of Chi-square is more than the table value therefore, Null hypothesis is rejected. This revealed that reaction is not uniformly distributed in the 5 point scale. More load is on disagree therefore most of the students agree with the statement "I face attention splash in e-learning for Equivalence Relations".

Statement 36: I know Education based websites for Equivalence Relations.

***Polarity*:** Positive

Responses of the students in percentage is given in Table 113 and Fig. 36.

10% of the students strongly agree with the statement "I know Education based websites for Equivalence Relations".

75% of the students agree with the statement "I know Education based websites for Equivalence Relations".

10% of the students not decided with the statement "I know Education based websites for Equivalence Relations".

5% of the students disagree with the statement "I know Education based websites for Equivalence Relations".

Table 113: Students response in percentage.

Points	*Response*	*Response percent*	*Response count*
5	Strongly agree	10%	2
4	Agree	75%	15
3	Not decided	10%	2
2	Disagree	5%	1
1	Strongly disagree	0%	0

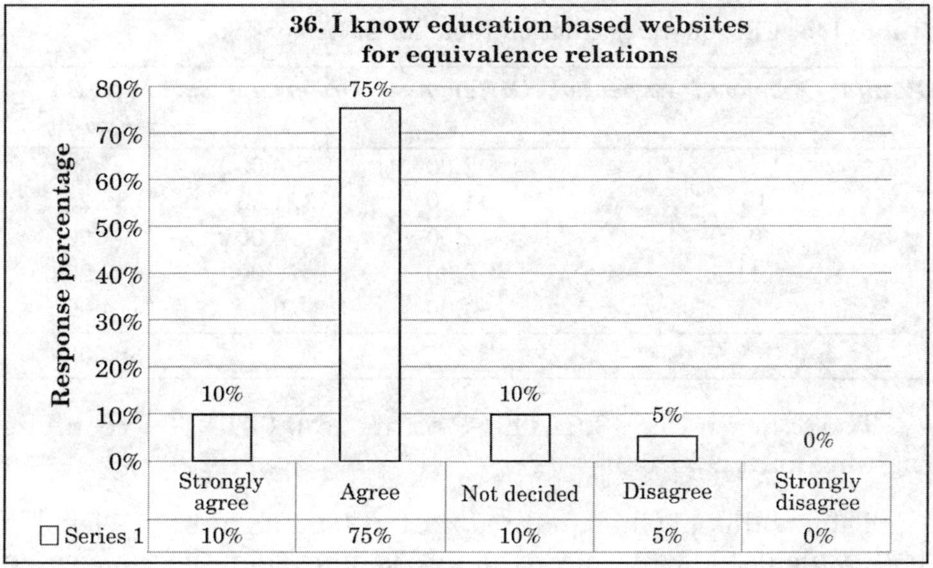

Fig. 36: Graphical representation of statement 36 in percentage.

0.00% of the students strongly disagree with the statement "I know Education based websites for Equivalence Relations".

Data analysis using Chi-square are given in Table 114 and Table 115.

H_0: Response is uniformly distributed in the 5 point scale

H_1: Response is not uniformly distributed in the 5 point scale

Table 114: Chi-square table for statement 36.

Points	*Observed frequency*	*Expected frequency*
5	2	4
4	15	4
3	2	4
2	1	4
1	0	4

Table 115: Chi-square calculation table for statement 36.

Points	*Observed*	*Expected*	*Difference*	*Difference sq.*	*Diff. Sq. / Exp fr.*
5	2	4	–2.00	4.00	1.00
4	15	4	11.00	121.00	30.25
3	2	4	–2.00	4.00	1.00
2	1	4	–3.00	9.00	2.25
1	0	4	–4.00	16.00	4.00
					38.500

The x^2 value is 38.5. The P-value is 0.001. The result is significant at $p = \leq 0.05$

Table value of Chi-square at 4 df at 0.05 significance level is 9.49. Calculated value of Chi-square is more than the table value therefore, Null hypothesis is rejected. This revealed that reaction is not uniformly distributed in the 5 point scale. More load is on **agree** therefore most of the students **agreed** with the statement "I know Education based websites for Equivalence Relations".

Statement 37: I face problem in learning Equivalence Relations through e-learning material.

***Polarity*:** Negative

Responses of the students in percentage is given in Table 116 and Fig. 37.

Table 116: Students response in percentage.

Points	*Response*	*Response percent*	*Response count*
1	Strongly agree	5%	1
2	Agree	25%	5
3	Not decided	15%	3
4	Disagree	55%	11
5	Strongly disagree	0%	0

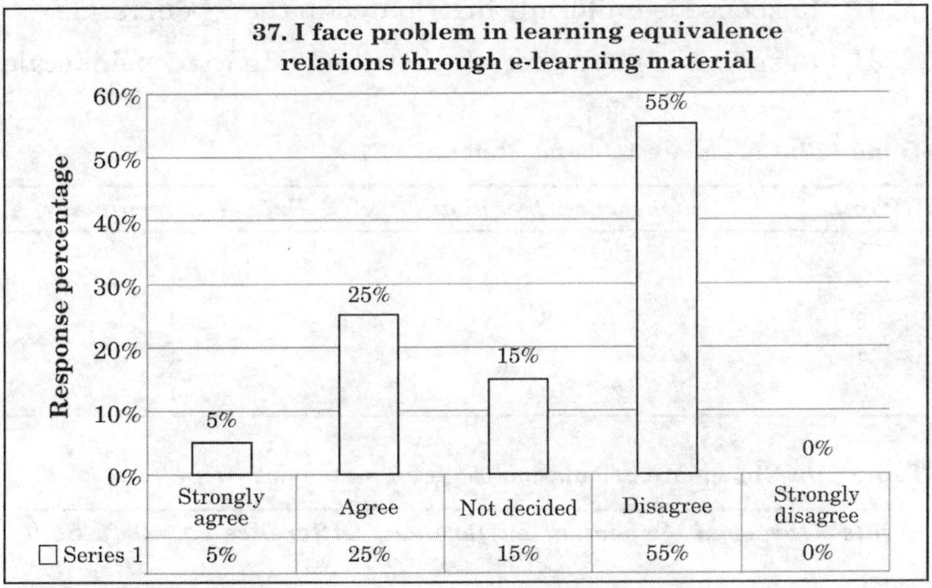

Fig. 37: Graphical representation of statement 37 in percentage.

5% of the students strongly agree with the statement "I face problem in learning Equivalence Relations through e-learning material".

25% of the students agree with the statement "I face problem in learning Equivalence Relations through e-learning material".

15% of the students not decided with the statement "I face problem in learning Equivalence Relations through e-learning material".

55% of the students disagree with the statement "I face problem in learning Equivalence Relations through e-learning material".

0.00% of the students strongly disagree with the statement "I face problem in learning Equivalence Relations through e-learning material".

Data analysis using Chi-square are given in Table 117 and Table 118.

H_0 : Response is uniformly distributed in the 5 point scale

H_1 : Response is not uniformly distributed in the 5 point scale

Table 117: Chi-square table for statement 37.

Points	*Observed frequency*	*Expected frequency*
1	1	4
2	5	4
3	3	4
4	11	4
5	0	4

Table 118: Chi-square calculation table for statement 37.

Points	*Observed*	*Expected*	*Difference*	*Difference sq.*	*Diff. Sq. / Exp fr.*
1	1	4	–3.00	9.00	2.25
2	5	4	1.00	1.00	0.25
3	3	4	–1.00	1.00	0.25
4	11	4	7.00	49.00	12.25
5	0	4	–4.00	16.00	4.00
					19.000

The x^2 value is 19. The P-value is 0.001. The result is significant at $p = \leq 0.05$

Table value of Chi-square at 4 df at 0.05 significance level is 9.49. Calculated value of Chi-square is more than the table value therefore, Null hypothesis is rejected. This revealed that reaction is not uniformly distributed in the 5 point scale. More load is on **disagree** therefore most of the students **disagreed** with the statement "I face problem in learning Equivalence Relations through e-learning material".

Statement 38: I get boredom in e-learning on learning Equivalence Relations.

***Polarity*:** Negative

Responses of the students in percentage in Table 119 and Fig. 38.

Table 119: Students response in percentage.

Points	***Response***	***Response percent***	***Response count***
1	Strongly agree	0%	0
2	Agree	15%	3
3	Not decided	30%	6
4	Disagree	55%	11
5	Strongly disagree	0%	0

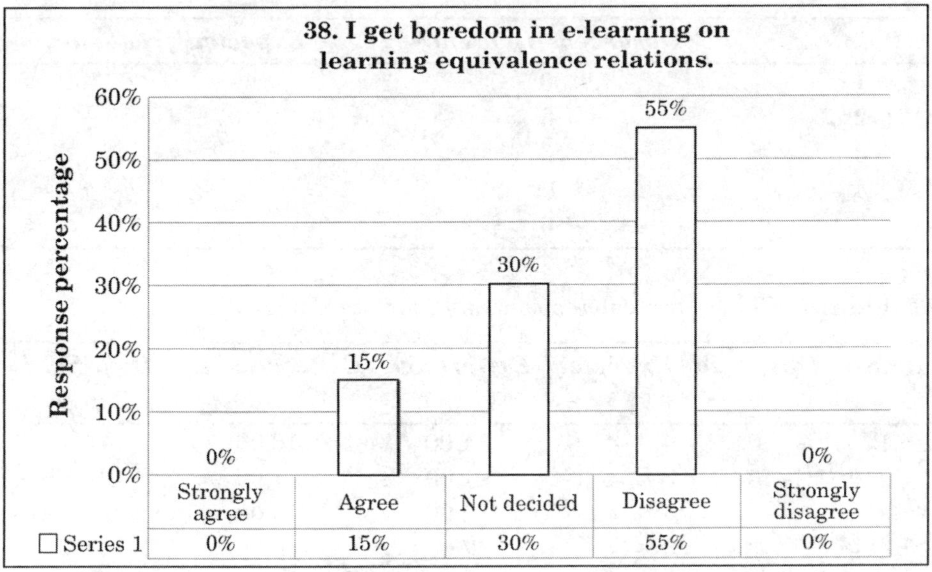

Fig. 38: Graphical representation of statement 38 in percentage.

0% of the students strongly agree with the statement "I get boredom in e-learning on learning Equivalence Relations".

15% of the students agree with the statement "I get boredom in e-learning on learning Equivalence Relations".

30% of the students not decided with the statement "I get boredom in e-learning on learning Equivalence Relations".

55% of the students disagree with the statement "I get boredom in e-learning on learning Equivalence Relations".

0.00% of the students strongly disagree with the statement "I get boredom in e-learning on learning Equivalence Relations".

Data analysis using Chi-square are given in Table 120 and Table 121.

H_0 : Response is uniformly distributed in the 5 point scale

H_1 : Response is not uniformly distributed in the 5 point scale

Table 120: Chi-square table for statement 38.

Points	*Observed frequency*	*Expected frequency*
1	0	4
2	3	4
3	6	4
4	11	4
5	0	4

Table 121: Chi-square calculation table for statement 38.

Points	*Observed*	*Expected*	*Difference*	*Difference sq.*	*Diff. Sq. / Exp fr.*
1	0	4	–4.00	16.00	4.00
2	3	4	–1.00	1.00	0.25
3	6	4	2.00	4.00	1.00
4	11	4	7.00	49.00	12.25
5	0	4	–4.00	16.00	4.00
					21.500

The x^2 value is 21.5. The P-value is 0.001. The result is significant at $p = \leq 0.05$

Table value of Chi-square at 4 df at 0.05 significance level is 9.49. Calculated value of Chi-square is more than the table value therefore, Null hypothesis is rejected. This revealed that reaction is not uniformly distributed in the 5 point scale. More load is on **disagree** therefore most of the students **disagreed** with the statement "I get boredom in e-learning on learning Equivalence Relations".

Statement 39: I need teacher's help while learning Equivalence Relations through e-learning.

***Polarity*:** Negative

Responses of the students in percentage is given in Table 122 and Fig. 39.

Table 122: Students response in percentage.

Points	*Response*	*Response percent*	*Response count*
1	Strongly agree	15%	3
2	Agree	45%	9
3	Not decided	0%	0
4	Disagree	40%	8
5	Strongly disagree	0%	0

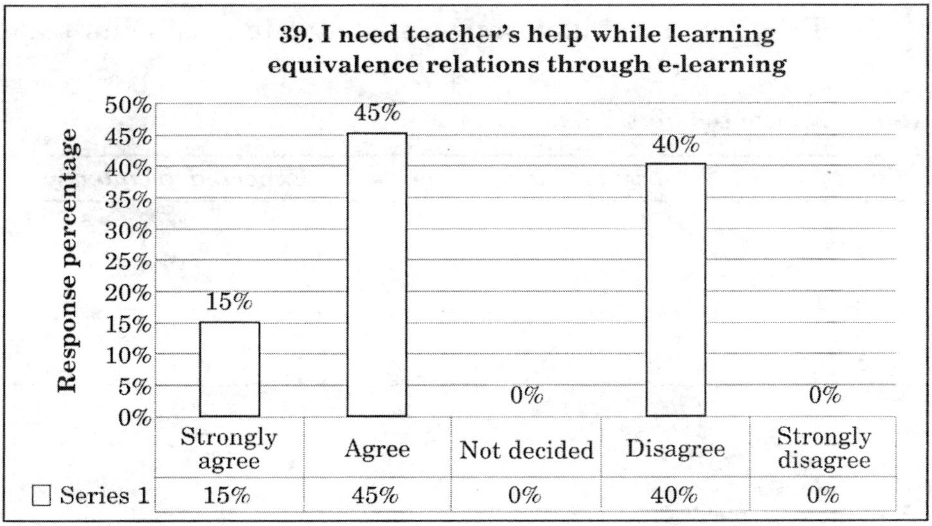

Fig. 39: Graphical representation of statement 39 in percentage.

15% of the students strongly agree with the statement “I need teacher’s help while learning Equivalence Relations through e-learning”.

45% of the students agree with the statement “I need teacher’s help while learning Equivalence Relations through e-learning”.

0% of the students not decided with the statement “I need teacher’s help while learning Equivalence Relations through e-learning”.

40% of the students disagree with the statement “I need teacher’s help while learning Equivalence Relations through e-learning”.

0.00% of the students strongly disagree with the statement “I need teacher’s help while learning Equivalence Relations through e-learning”.

Data analysis using Chi-square are given in Table 123 and Table 124.

H_0 : Response is uniformly distributed in the 5 point scale

H_1 : Response is not uniformly distributed in the 5 point scale

Table 123: Chi-square table for statement 39.

Points	*Observed frequency*	*Expected frequency*
1	3	4
2	9	4
3	0	4
4	8	4
5	0	4

Table 124: Chi-square calculation table for statement 39.

Points	*Observed*	*Expected*	*Difference*	*Difference sq.*	*Diff. Sq. / Exp fr.*
1	3	4	–1.00	1.00	0.25
2	9	4	5.00	25.00	6.25
3	0	4	–4.00	16.00	4.00
4	8	4	4.00	16.00	4.00
5	0	4	–4.00	16.00	4.00
					18.500

The x^2 value is 18.5. The P-value is 0.001. The result is significant at $p = \leq 0.05$

Table value of Chi-square at 4 df at 0.05 significance level is 9.49. Calculated value of Chi-square is more than the table value therefore, Null hypothesis is rejected. This revealed that reaction is not uniformly distributed in the 5 point scale. More load is on agree therefore most of the students agree with the statement "I need teacher's help while learning Equivalence Relations through e-learning".

Statement 40: I enjoyed e-learning as I can watch video in the topic Equivalence Relations.

***Polarity*:** Positive

Responses of the students in percentage is given in Table 125 and Fig. 40.

Table 125: Students response in percentage.

Points	*Response*	*Response percent*	*Response count*
5	Strongly agree	25%	5
4	Agree	60%	12
3	Not decided	10%	2
2	Disagree	0%	0
1	Strongly disagree	5%	1

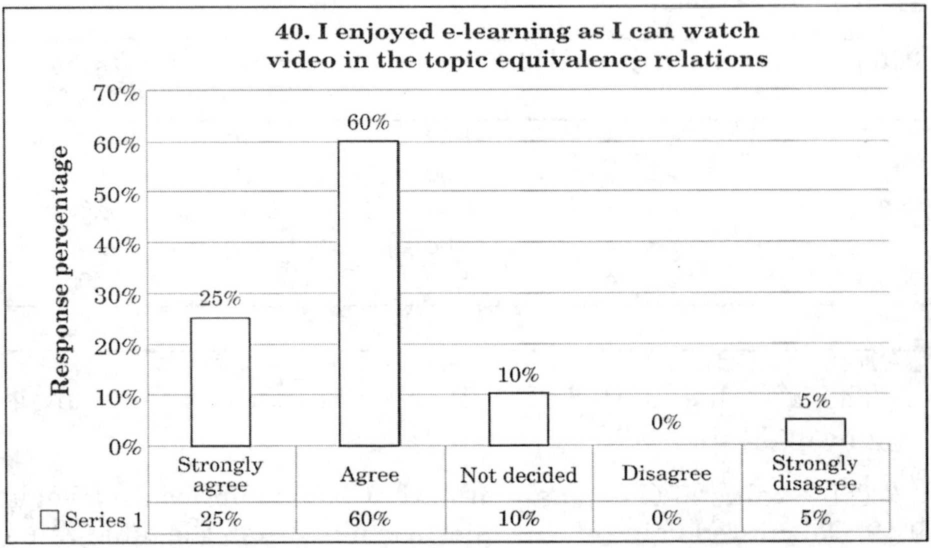

Fig. 40: Graphical representation of statement 40 in percentage.

25% of the students strongly agree with the statement "I enjoyed e-learning as I can watch video in the topic Equivalence Relations".

60% of the students agree with the statement "I enjoyed e-learning as I can watch video in the topic Equivalence Relations".

10% of the students not decided with the statement "I enjoyed e-learning as I can watch video in the topic Equivalence Relations".

0% of the students disagree with the statement "I enjoyed e-learning as I can watch video in the topic Equivalence Relations".

5% of the students strongly disagree with the statement "I enjoyed e-learning as I can watch video in the topic Equivalence Relations".

Data analysis using Chi-square are given in Table 126 and Table 127.

H_0 : Response is uniformly distributed in the 5 point scale

H_1 : Response is not uniformly distributed in the 5 point scale

Table 126: Chi-square table for statement 40.

Points	*Observed frequency*	*Expected frequency*
5	5	4
4	12	4
3	2	4
2	0	4
1	1	4

Table 127: Chi-square calculation table for statement 40.

Points	*Observed*	*Expected*	*Difference*	*Difference sq.*	*Diff. Sq. / Exp fr.*
5	5	4	1.00	1.00	0.25
4	12	4	8.00	64.00	16.00
3	2	4	–2.00	4.00	1.00
2	0	4	–4.00	16.00	4.00
1	1	4	–3.00	9.00	2.25
					23.500

The x^2 value is 23.5. The P-value is 0.001. The result is significant at $p = \leq 0.05$

Table value of Chi-square at 4 df at 0.05 significance level is 9.49. Calculated value of Chi-square is more than the table value therefore, Null hypothesis is rejected. This revealed that reaction is not uniformly distributed in the 5 point scale. More load is on **agree** therefore most of the students **agreed** with the statement "I enjoyed e-learning as I can watch video in the topic Equivalence Relations".

14.6.3. *Comprehensive analysis of reaction scale*

Data analysis of responses of students is presented through Table 128.

Tabulated Value of χ^2 at 4 df at .05 level is 9.49.

Table 128: Comprehensive analysis of responses on reaction scale given by experimental group students.

Sl. no.	*Statement*	*Polarity*	*Calculated value of χ^2 at 4 df .05 level*	*Significant difference*	*Maximum load*	*Points*
1	I like to learn Recurrence relations through e-learning.	Positive	32	Yes	Agree (Positive side)	4
2	I can learn Recurrence relations without Teacher and only through e-learning material.	Positive	15.5	Yes	Agree (Positive side)	4
3	I face problem in computer based learning at Recurrence relations.	Negative	12.5	Yes	Disagree (Negative side)	4
4	I face problem in answering Google's Quiz online test in Recurrence relations.	Negative	19.5	Yes	Disagree (Negative side)	4
5	I face attention splash in e-learning for Recucurrence relations.	Negative	19.5	Yes	Agree (Negative side)	2
6	I know Education based websites for Recurrence relations.	Positive	53	Yes	Agree (Positive side)	4
7	I face problem in learning Recurrence Relations through e-learning material.	Negative	23	Yes	Disagree (Negative side)	4
8	I get boredom in e-learning on learning Recurrence Relations.	Negative	11.5	Yes	Disagree (Negative side)	4
9	I need teacher's help while learning Recurr-	Negative	13.5	Yes	Agree (Negative side)	2

Table 128: (*Contd...*)

Table 128: (*Contd...*)

Sl. no.	*Statement*	*Polarity*	*Calculated value of χ^2 at 4 df .05 level*	*Significant difference*	*Maximum load*	*Points*
	ence Relations through e-learning.					
10	I enjoyed e-learning as I can watch video in the topic Recurrence Relations.	Positive	27.5	Yes	Agree (Positive side)	4
11	I like to learn Permutations through e-learning.	Positive	16	Yes	Agree (Positive side)	4
12	I can learn Permutations without Teacher and only through e-learning material.	Positive	22.5	Yes	Agree (Positive side)	4
13	I face problem in computer based learning at Permutations.	Negative	9.00	Yes	Disagree (Negative side)	4
14	I face problem in answering Google's Quiz online test in Permutations.	Negative	26.00	Yes	Disagree (Negative side)	4
15	I face attention splash in e-learning for Permutations.	Negative	11.00	Yes	Disagree (Negative side)	4
16	I know Education based websites for Permutations.	Positive	29.5	Yes	Agree (Positive side)	4
17	I face problem in learning Permutations through e-learning material.	Negative	32.00	Yes	Disagree (Negative side)	4

Table 128: (*Contd...*)

Table 128: *(Contd...)*

Sl. no.	*Statement*	*Polarity*	*Calculated value of χ^2 at 4 df .05 level*	*Significant difference*	*Maximum load*	*Points*
18	I get boredom in e-learning on learning Permutations.	Negative	11.00	Yes	Equal load on agree and Disagree (Negative side)	(3) average of 2 and 4
19	I need teacher's help while learning Permutations through e-learning.	Negative	10.50	Yes	Disagree (Negative side)	4
20	I enjoyed e-learning as I can watch video in the topic Permutations	Positive	22.5	Yes	Agree (Positive side)	4
21	I like to learn Groups through e-learning.	Positive	26.5	Yes	Agree (Positive side)	4
22	I can learn Groups without Teacher and only through e-learning material.	Positive	8.5	No	Agree (Positive side)	–
23	I face problem in computer based learning at Groups.	Negative	11.50	Yes	Disagree (Negative side)	4
24	I face problem in answering Google's Quiz online test in Groups.	Negative	23.50	Yes	Disagree (Negative side)	4
25	I face attention splash in e-learning for Groups.	Negative	12.50	Yes	Disagree (Negative side)	4
26	I know Education based websites for Groups.	Positive	27.5	Yes	Agree (Positive side)	4

Table 128: *(Contd...)*

Table 128: (*Contd...*)

Sl. no.	*Statement*	*Polarity*	*Calculated value of χ^2 at 4 df .05 level*	*Significant difference*	*Maximum load*	*Points*
27	I face problem in learning Groups through e-learning material.	Negative	17.50	Yes	Disagree (Negative side)	4
28	I get boredom in e-learning on learning Groups.	Negative	21.50	Yes	Equal load on agree and Disagree (Negative side)	(3) average of 2 and 4
29	I need teacher's help while learning Groups through e-learning.	Negative	16.50	Yes	Agree (Negative side)	2
30	I enjoyed e-learning as I can watch video in the topic Groups	Positive	13.5	Yes	Agree (Positive side)	4
31	I like to learn Equivalence Relations through e-learning.	Positive	38.5	Yes	Agree (Positive side)	4
32	I can learn Equivalence Relations without Teacher and only through e-learning material.	Positive	16.5	Yes	Agree (Positive side)	4
33	I face problem in computer based learning at Equivalence Relations.	Negative	23.50	Yes	Disagree (Negative side)	4
34	I face problem in answering Google's Quiz online test in Equivalence Relations.	Negative	28.50	Yes	Disagree (Negative side)	4

Tablc 128: (*Contd...*)

Table 128: (*Contd...*)

Sl. no.	*Statement*	*Polarity*	*Calculated value of χ^2 at 4 df .05 level*	*Significant difference*	*Maximum load*	*Points*
35	I face attention splash in e-learning for Equivalence Relations.	Negative	17.50	Yes	Disagree (Negative side)	4
36	I know Education based websites for Equivalence Relations.	Positive	38.5	Yes	Agree (Positive side)	4
37	I face problem in learning Equivalence Relations through e-learning material.	Negative	19.00	Yes	Disagree (Negative side)	4
38	I get boredom in e-learning on learning Equivalence Relations.	Negative	21.50	Yes	Disagree (Negative side)	4
39	I need teacher's help while learning Equivalence Relations through e-learning.	Negative	18.50	Yes	Agree (Negative side)	2
40	I enjoyed e-learning as I can watch video in the topic Equivalence Relations.	Positive	23.5	Yes	Agree (Positive side)	4

Section 5

15. FINDINGS FROM THE ANALYSIS OF REACTION SCALE

Experimental Group

Out of total ***forty*** statements bearing positive as well as negative nature, the computed Chi-square values of ***thirty nine*** statements were found to have statistically significant higher values than the Chi-square table values 9.49 at 4 degrees of freedom and at .05 level of significance which shows that there was significant difference between the observed and expected frequencies therefore null hypothesis is not rejected. This reveals that reaction is uniformly distributed in the 5 point scale.

The computed Chi-square values of ***two*** statements were found to have statistically significant higher values than the Chi-square table values 9.49 at 4 degrees of freedom and at .05 level of significance which shows that there was significant difference between the observed and expected frequencies therefore null hypothesis is rejected. This reveals that reaction is not uniformly distributed in the 5 point scale. Students were found to have neutral attitude (equal load of positive as well as negative values) towards the ***two*** statements.

The computed Chi-square values of ***four*** statements were found to have statistically significant higher values than the Chi-square table values 9.49 at 4 degrees of freedom and at .05 level of significance which shows that there was significant difference between the observed and expected frequencies therefore null hypothesis is rejected. This reveals that reaction is not uniformly distributed in the 5 point scale. Students were found to have negative attitude towards the ***four*** statements.

The computed Chi-square values of ***thirty three*** statements was found to have statistically significant higher values than the Chi-square table values 9.49 at 4 degrees of freedom and at .05 level of significance which shows that there was significant difference between the observed and expected frequencies therefore null hypothesis is rejected. This reveals that reaction is not uniformly distributed in the 5 point scale. Students were found to have positive attitude towards the ***thirty three*** statements.

16. INTERPRETATION OF RESULTS

16.1. Interpretation of ANOVA

It was found that there was significant difference between the mean achievement score of Experimental Group students and the mean achievement score of Control Group students therefore the hypothesis "There will be no significant difference in the achievement scores of experimental group students and control group students" is rejected. We can conclude that students learnt well using e-learning material than conventional method; e-learning is effective than conventional method. We can also safely say that e-learning material can be used as a substitution for conventional learning wherever and whenever necessary.

16.2. Interpretation of Result of Reaction Scale

Findings from the analysis of reaction scale for experimental group indicate that out of **forty** statements the computed Chi-square values of **thirty three** statements were found to have statistically significant higher values than the tabulated value and the students were found to have **positive reaction** and **favorable attitude** towards the statements carrying such **higher values.** That is, for more than **50%** of the statements they responded **positively**.

From the above paragraphs we can conclude that students liked learning discrete mathematics through e-learning material and investigators observed that the students liked the concept of e-learning and enjoyed while learning. We can conclude that e-learning is one of the effective methods to learn discrete mathematics.

17. E-LEARNING IMPLEMENTED IN OTHER STATES FOR FURTHER ANALYSIS

E-learning material is further analyzed in other states of India for further understanding of the material developed. Therefore investigators visited other parts of India other than randomly selected colleges for the study. E-learning material was implemented and effectiveness was tested.

17.1. Testing the Effectiveness of Developed E-learning Material in South India, Madurai

Step 1

Investigators contacted the Principal, Dr. D. Pandiaraja and asked for permission to implement the material in their esteemed college. He agreed and gave us permission and implementation date was fixed accordingly.

Step 2

As per the schedule investigators visited Thiagarajar College of Arts and Science on 20.02.2018 and implemented the developed e-learning material in four topics. After completion of implementation students were given questionnaire in four topics to test the effectiveness of the developed e-learning material. Questionnaires were analyzed and the analyzed data (result) is displayed in **Tables 129, 130, 131 and 132**. Investigators also observed the students and two students were interviewed about the developed e-learning material. Mostly students liked the material and enthusiastically participated throughout the experiment. They asked questions whenever they have doubts and even helped other students in understanding. Students asked many questions like why we developed only in these topics and not in other topics. General questions were asked in interview to know their opinion about developed material. Student's answers were positive and they liked to learn using e-learning material. Investigators also noted that there were one or two students who should be monitored

Table 129: Analysis of Equivalence Relation Thiagarajar College of Arts and Science, Madurai, Tamil Nadu.

Sl. no.	*Questions*	*No. of Yes*	*% of Yes*	*No. of No*	*% of No*	*Colu-mn 1*	*Colu-mn 2*
1	Do you like to learn Equivalence Relation through e-learning?	47	100.00%	0	0%		
2	Will you learn Equivalence Relation without Teacher through Computer?	41	87%	6	13.00%		
3	Is there any problem in computer based learning at Equivalence Relation?	10	21%	37	79.00%		
4	Are you face any problem in Google's Quiz online test?	4	9%	43	91.00%		
5	Any chance to face attention splash in e-learning for Equivalence Relation?	9	19%	38	81.00%		
6	Do you know any Education based websites for Equivalence Relation?	9	19%	38	81.00%		
7	Did you face any problem in Equivalence Relation in e-learning?	20	43%	27	57.00%		
8	Which instrument you like in e-learning? (like TV, Radio, Computer)	No. of like computer	% of like computer	No. of like TV	% of like TV		
		40	85%	7	15%		
9	Which learning software is suitable for Equivalence Relation in computer? (PPT, Video)	No. of like PPT	% of like PPT	No. of like video	% of like video	No. of like both	% of like both
		34	72%	7	14%	6	14%
10	Which one of the test you like online test or offline test for Equivalence Relation?	No. of like online	% of like online	No. of like offline	% of like offline		
		21	45%	26	55%		

Table 130: Analysis of Groups, Thiagarajar College of Arts and Science, Madurai, Tamil Nadu.

Sl. no.	*Questions*	*No. of Yes*	*% of Yes*	*No. of No*	*% of No*	*Column 1*	*Column 2*
1	Do you like to learn Groups through e-learning?	24	96.00%	1	4%		
2	Will you learn Groups without Teacher through Computer?	22	88%	3	12.00%		
3	Is there any problem in computer based learning at Groups?	1	4%	24	96.00%		
4	Are you face any problem in Google's Quiz online test?	2	8%	23	92.00%		
5	Any chance to face attention splash in e-learning for Groups?	1	4%	24	96.00%		
6	Do you know any Education based websites for Groups?	5	20%	20	80.00%		
7	Did you face any problem in Groups e-learning?	2	8%	23	92.00%		
8	Which instrument you like in e-learning? (like TV, Radio, Computer)	No. of like computer	% of like computer	No. of like TV	% of like TV		
		19	76%	6	24%		
9	Which learning software is suitable for Groups in computer? (PPT, Video)	No. of like PPT	% of like PPT	No. of like video	% of like video	No. of like both	% of like both
		15	60%	7	28%	5	12%
10	Which one of the test you like online test or offline test for Groups?	No. of like online	% of like online	No. of like offline	% of like offline		
		12	48%	6	52%		

Table 131: Analysis of Permutations, Thiagarajar College of Arts and Science, Madurai, Tamil Nadu.

Sl. no.	*Questions*	*No. of Yes*	*% of Yes*	*No. of No*	*% of No*	*Column 1*	*Column 2*
1	Do you like to learn Permutations through e-learning?	43	97%	1	3%		
2	Will you learn Permutations without Teacher through Computer?	38	86.00%	6	14%		
3	Is there any problem in computer based learning at Permutations?	7	16%	37	84.00%		
4	Are you face any problem in Google's Quiz online test?	13	30%	31	70.00%		
5	Any chance to face attention splash in e-learning for Permutations?	15	35%	29	65.00%		
6	Do you know any Education based websites for Permutations?	11	25%	33	75.00%		
7	Did you face any problem in Permutations e-learning?	20	45%	24	55%		
8	Which instrument you like in e-learning? (like TV, Radio, Computer)	No. of like computer	% of like computer	No. of like TV	% of like TV		
		41	93%	3	7%		
9	Which learning software is suitable for Permutations in computer? (PPT, Video)	No. of like PPT	% of like PPT	No. of like video	% of like video	No. of like both	% of like both
		31	70%	7	15%	6	15%
10	Which one of the test you like online test or offline test for Permutations?	No. of like online	% of like online	No. of like offline	% of like offline		
		21	47%	23	53%		

Table 132: Analysis of Recurrence Relations, Thiagarajar College of Arts and Science, Madurai, Tamil Nadu.

Sl. no.	*Questions*	*No. of Yes*	*% of Yes*	*No. of No*	*% of No*	*Column 7*	*Column 8*
1	Do you like to learn Recurrence relations through e-learning?	37	100.00%	0	0%		
2	Will you learn Recurrence relations without Teacher through Computer?	32	86%	5	14.00%		
3	Is there any problem in computer based learning at Recurrence relations?	5	13%	32	87.00%		
4	Are you face any problem in Google's Quiz online test?	8	22%	29	78.00%		
5	Any chance to face attention splash in e-learning for Recurrence relations?	3	8%	34	92.00%		
6	Do you know any Education based websites for Recurrence relations?	7	19%	30	81%		
7	Did you face any problem in Recurrence relations e-learning?	9	24%	28	76%		
8	Which instrument you like in e-learning? (like TV, Radio, Computer...)	No. of like computer	% of like computer	No. of like TV	% of like TV		
		35	95%	2	5%		
9	Which learning software is suitable for Recurrence relations in computer? (PPT, Video...)	No. of like PPT	% of like PPT	No. of like video	% of like video	No. of like both	% of like both
		26	70%	7	19%	4	11%
10	Which one of the test you like online test or offline test for Recurrence relations?	No. of like online	% of like online	No. of like offline	% of like offline		
		18	49%	19	51%		

continuously to ensure that they are learning or not otherwise they will engage themselves in other activities.

Step 3

Through observation it is noted that student's intrinsic motivation plays very important role in learning through this material. Self-motivated learners learn very easily through this material. There are very few students in the class who are not self-motivated there is quite possible that there can be attention splash after few hours of learning.

There were 47 B.Sc. Computer Science students in the selected class in Thiagarajar College of Arts and Science. Their online feedback was analyzed and its results are as follows.

17.1.1. *Topic equivalence relation*

The following (Table 129) is the analysis of the questionnaire result for the topic Equivalence Relation.

There were six positively oriented questions and four negatively oriented questions more load (more than 50% of students) is on **YES** in positively oriented questions (questions 1, 2, 8, 9) for positively oriented questions 6 and 10 more load in on **NO** side, more load (more than 50% of students) is on **NO** in negatively oriented questions (3, 4, 5, 7).

Analysis clearly shows that most of the students liked to learn the topic Equivalence Relation using e-learning material.

17.1.2. *Topic groups*

The following (Table 130) is the analysis of the questionnaire result for the topic groups.

There were six positively oriented questions and four negatively oriented questions more load (more than 50% of students) is on **YES** in positively oriented questions (questions 1, 2, 8, 9) for positively oriented questions 6 and 10 more load in on **NO** side, more load (more than 50% of students) is on **NO** in negatively oriented questions (3, 4, 5, 7).

Analysis clearly shows that most of the students liked to learn the topic Groups using e-learning material.

17.1.3. *Topic permutations*

The following (Table 131) is the analysis of the questionnaire result for the topic Permutations.

There were six positively oriented questions and four negatively oriented questions more load (more than 50% of students) is on **YES** in positively oriented questions (questions 1, 2, 8, 9) for positively oriented questions 6 and 10 more load in on **NO** side, more load (more than 50% of students) is on **NO** in negatively oriented questions (3, 4, 5, 7).

Analysis clearly shows that most of the students liked to learn the topic Permutations using e-learning material.

17.1.4. *Topic recurrence relations*

The following (Table 132) is the analysis of the questionnaire result for the topic Recurrence Relations.

There were six positively oriented questions and four negatively oriented questions more load (more than 50% of students) is on **YES** in positively oriented questions (questions 1, 2, 8, 9) for positively oriented questions 6 and 10 more load in on **NO** side, more load (more than 50% of students) is on **NO** in negatively oriented questions (3, 4, 5, 7).

Analysis clearly shows that most of the students liked to learn the topic Recurrence relations using e-learning material.

17.2. Testing the Effectiveness of Developed E-learning Material in West India, Vadodara

Step 1

Investigators contacted the Coordinator, Dr. Mandira Sikdar, Navrachana University B.Sc. Programme and asked for permission to implement the material in their esteemed University. She agreed and gave us permission and implementation date was fixed accordingly.

Step 2

As per the schedule investigators visited Navrachana University from 8th January to 12th January 2018 and implemented the developed e-learning material in four topics. After completion of implementation students were given questionnaire in four topics to test the effectiveness of the developed e-learning material. Questionnaires were analyzed and the analyzed data (result) is displayed in **Tables 133, 134, 135 and 136**. Investigators also observed the students and three students were interviewed about the developed e-learning material. Mostly students liked the material and enthusiastically participated throughout the experiment. They asked questions whenever they have doubts and even helped other students in understanding. Students asked many questions like whether the results of the experiment will be communicated to them, whether they can refer the material again and whether they can copy the material in their pen drive. Materials were loaded to the computer through the server. Generic questions were asked in interview to know their opinion about developed material and their problems in learning through this material. Student's answers were positive and they liked to learn using e-learning material. They wanted more such material so that it will help them to consolidate their learning in classroom method and will help in revision. Sometimes they have problems in learning in classroom teaching when they were absent for previous class. If they have such material with them they can overcome this learning gap. They also suggested some ideas to make the material more

Table 133: Analysis of Equivalence Relation Navrachana University, Vadodara, Gujarat.

Sl. no.	***Questions***	***No. of Yes***	***% of Yes***	***No. of No***	***% of No***	***Colu-mn 1***	***Colu-mn 2***
1	Do you like to learn Equivalence Relation through e-learning?	9	64%	5	36%		
2	Will you learn Equivalence Relation without Teacher through Computer?	12	85.00%	2	15.00%		
3	Is there any problem in computer based learning at Equivalence Relation?	0	0%	14	100.00%		
4	Are you face any problem in Google's Quiz online test?	0	0%	14	100.00%		
5	Any chance to face attention splash in e-learning for Equivalence Relation?	2	14%	12	86.00%		
6	Do you know any Education based websites for Equivalence Relation?	1	7%	13	93.00%		
7	Did you face any problem in Equivalence Relation in e-learning?	0	0%	14	100.00%		
8	Which instrument you like in e-learning? (like TV, Radio, Computer...)	No. of like computer	% of like computer	No. of like TV	% of like TV		
		12	95%	2	5%		
9	Which learning software is suitable for Equivalence relation in computer? (PPT, Video...)	No. of like PPT	% of like PPT	No. of like video	% of like video	No. of like both	% of like both
		8	30%	4	40%	2	30%
10	Which one of the test you like online test or offline test for Equivalence relation?	No. of like online	% of like online	No. of like offline	% of like offline		
		14	70%	6	30%		

Table 134: Analysis of Groups, Navrachana University, Vadodara, Gujarat.

Sl. no.	*Questions*	*No. of Yes*	*% of Yes*	*No. of No*	*% of No*	*Colu-mn 7*	*Colu-mn 8*
1	Do you like to learn Groups through e-learning?	8	57.00%	6	43%		
2	Will you learn Groups without Teacher through Computer?	11	78%	3	22.00%		
3	Is there any problem in computer based learning at Groups?	1	7%	13	93.00%		
4	Are you face any problem in Google's Quiz online test?	1	7%	13	93.00%		
5	Any chance to face attention splash in e-learning for Groups?	1	7%	13	93.00%		
6	Do you know any Education based websites for Groups?	0	0%	14	100.00%		
7	Did you face any problem in Groups e-learning?	0	0%	14	100.00%		
8	Which instrument you like in e-learning? (like TV, Radio, Computer...)	No. of like comp-uter	% of like comp-uter	No. of like TV	% of like TV		
		12	95%	2	10%		
9	Which learning software is suitable for Groups in computer? (PPT, Video...)	No. of like PPT	% of like PPT	No. of like video	% of like video	No. of like both	% of like both
		8	30%	3	15%	5	25%
10	Which one of the test you like online test or offline test for Groups?	No. of like online	% of like online	No. of like offline	% of like offline		
		14	70%	6	30%		

Table 135: Analysis of Permutations, Navrachana University, Vadodara, Gujarat.

Sl. no.	*Questions*	*No. of Yes*	*% of Yes*	*No. of No*	*% of No*	*Colu-mn 7*	*Colu-mn 8*
1	Do you like to learn Permutations through e-learning?	9	64%	5	36%		
2	Will you learn Permutations without Teacher through Computer?	12	85.00%	2	15.00%		
3	Is there any problem in computer based learning at Permutations?	0	0%	14	100.00%		
4	Are you face any problem in Google's Quiz online test?	0	0%	14	100.00%		
5	Any chance to face attention splash in e-learning for Permutations?	2	14%	12	86.00%		
6	Do you know any Education based websites for Permutations?	1	7%	13	93.00%		
7	Did you face any problem in Permutations e-learning?	0	0%	14	100.00%		
8	Which instrument you like in e-learning? (like TV, Radio, Computer...)	No. of like computer	% of like computer	No. of like TV	% of like TV		
		12	95%	2	5%		
9	Which learning software is suitable for Permutations in computer? (PPT, Video...)	No. of like PPT	% of like PPT	No. of like video	% of like video	No. of like both	% of like both
		8	30%	4	40%	2	30%
10	Which one of the test you like online test or offline test for Permutations?	No. of like online	% of like online	No. of like offline	% of like offline		
		14	70%	0	30%		

Table 136: Analysis of Recurrence Relations, Navrachana University, Vadodara, Gujarat.

Sl. no.	*Questions*	*No. of Yes*	*% of Yes*	*No. of No*	*% of No*	*Column 7*	*Column 8*
1	Do you like to learn Recurrence relations through e-learning?	8	57.00%	6	43%		
2	Will you learn Recurrence relations without Teacher through Computer?	11	78%	3	22.00%		
3	Is there any problem in computer based learning at Recurrence relations?	1	7%	13	93.00%		
4	Are you face any problem in Google's Quiz online test?	1	7%	13	93.00%		
5	Any chance to face attention splash in e-learning for Recurrence relations?	1	7%	13	93.00%		
6	Do you know any Education based websites for Recurrence relations?	0	0%	14	100.00%		
7	Did you face any problem in Recurrence relations e-learning?	0	0%	14	100.00%		
8	Which instrument you like in e-learning? (like TV, Radio, Computer...)	No. of like computer	% of like computer	No. of like TV	% of like TV		
		12	95%	2	10%		
9	Which learning software is suitable for Recurrence relations in computer? (PPT, Video...)	No. of like PPT	% of like PPT	No. of like video	% of like video	No. of like both	% of like both
		8	30%	3	15%	5	25%
10	Which one of the test you like online test or offline test for Recurrence relations?	No. of like online	% of like online	No. of like offline	% of like offline		
		14	70%	6	30%		

interesting like adding some skit related to the topics, mathematics games and also to add some riddles. Some students also suggested to add more pictures to make the topic more understandable.

Step 3

Through observation it is noted that student's intrinsic motivation plays very important role in learning through this material. Self-motivated learners learn very easily through this material. There are very few students in the class who are not self-motivated there is quite possible that there can be attention splash after few hours of learning.

There were 14 B.Sc. Mathematics students in the selected class in Navrachana University. Their online feedback was analyzed and its results are as follows.

17.2.1. *Topic equivalence relation*

The following (Table 133) is the analysis of the questionnaire result for the topic Equivalence Relation.

There were six positively oriented questions and four negatively oriented questions more load (more than 50% of students) is on **YES** in positively oriented questions (questions 1, 2, 8, 9, 10) for positively oriented questions 6 more load in on **NO** side, more load (more than 50% of students) is on **NO** in negatively oriented questions (3, 4, 5, 7).

Analysis clearly shows that most of the students liked to learn the topic Equivalence Relation using e-learning material.

17.2.2. *Topic groups*

The following (Table 134) is the analysis of the questionnaire result for the topic Groups.

There were six positively oriented questions and four negatively oriented questions more load (more than 50% of students) is on **YES** in positively oriented questions (questions 1, 2, 8, 9, 10) for positively oriented questions 6 more load in on **NO** side, more load (more than 50% of students) is on **NO** in negatively oriented questions (3, 4, 5, 7).

Analysis clearly shows that most of the students liked to learn the topic Groups using e learning material.

17.2.3. *Topic permutations*

The following (Table 135) is the analysis of the questionnaire result for the topic Permutations.

There were six positively oriented questions and four negatively oriented questions more load (more than 50% of students) is on **YES** in positively oriented questions (questions 1, 2, 8, 9, 10) for positively oriented questions 6 more load in on **NO** side, more load (more than 50% of students) is on **NO** in negatively oriented questions (3, 4, 5, 7).

Analysis clearly shows that most of the students liked to learn the topic Permutations using e-learning material.

17.2.4. *Topic recurrence relations*

The following (Table 136) is the analysis of the questionnaire result for the topic Recurrence Relations.

There were six positively oriented questions and four negatively oriented questions more load (more than 50% of students) is on **YES** in positively oriented questions (questions 1, 2, 8, 9, 10) for positively oriented questions 6 more load in on **NO** side, more load (more than 50% of students) is on **NO** in negatively oriented questions (3, 4, 5, 7).

Analysis clearly shows that most of the students liked to learn the topic Recurrence relations using e-learning material.

17.3. Testing the Effectiveness of Developed E-learning Material in North East India, RGU Arunachal Pradesh

Step 1

Investigators contacted the Coordinator, Mr. Firos A and HOD Dr. Ani Taggu, Department of Computer Science and Engineering Rajiv Gandhi University, Arunachal Pradesh and asked for permission to implement the material in their esteemed University. They agreed and gave us permission and implementation date was fixed accordingly.

Step 2

As per the schedule investigators visited Rajiv Gandhi University on 20th August 2018 and implemented the developed e-learning material in four topics. After completion of implementation students were given questionnaire in four topics to test the effectiveness of the developed e-learning material. Questionnaires were analyzed and the analyzed data (result) is displayed in **Tables 137, 138, 139 and 140**. Investigators also observed the students. The students consisted of mixed groups' seven students from M.C.A, eight students from Master of Technology in Computer Science and Engineering, three students from B.C.A. Mostly students liked the material and enthusiastically participated throughout the experiment. They asked questions whenever they have doubts and even helped other students in understanding. B.C.A. students took more time to complete the e-learning material and they were very enthusiastic in learning. One student was very interested in learning how we developed this material. He asked several questions like why we are conducting such experiments. Like him most of the students wanted to know the purpose of this experiment and what we will do after conducting this experiment. Few students took selfie with the investigators for their remembrance.

Table 137: Analysis of Equivalence Relation Rajiv Gandhi University, Arunachal Pradesh.

Sl. no.	*Questions*	*No. of Yes*	*% of Yes*	*No. of No*	*% of No*	*Colu-mn 1*	*Colu-mn 2*
1	Do you like to learn Equivalence Relation through e-learning?	18	100%	0	0%		
2	Will you learn Equivalence Relation without Teacher through Computer?	12	66.66%	6	33.33%		
3	Is there any problem in computer based learning at Equivalence Relation?	2	11.11%	16	88.88%		
4	Are you face any problem in Google's Quiz online test?	6	33.33%	12	66.66%		
5	Any chance to face attention splash in e-learning for Equivalence Relation?	2	14%	12	86.00%		
6	Do you know any Education based websites for Equivalence Relation?	5	27.77%	13	72.22%		
7	Did you face any problem in Equivalence Relation in e-learning?	3	16.66%	15	83.33%		
8	Which instrument you like in e-learning? (like TV, Radio, Computer...)	No. of like computer	% of like computer	No. of like TV	% of like TV		
		17	94.44%	1	5.55%		
9	Which learning software is suitable for Equivalence relation in computer? (PPT, Video...)	No. of like PPT	% of like PPT	No. of like video	% of like video	No. of like both	% of like both
		3	16.66%	4	22.22%	11	61.11%
10	Which one of the test you like online test or offline test for Equivalence relation?	No. of like online	% of like online	No. of like offline	% of like offline		
		6	33.33%	5	27.77%	7	38.88%

Table 138: Analysis of Groups, Rajiv Gandhi University, Arunachal Pradesh.

Sl. no.	*Questions*	*No. of Yes*	*% of Yes*	*No. of No*	*% of No*	*Colu-mn 7*	*Colu-mn 8*
1	Do you like to learn Groups through e-learning?	18	100%	0	0.00%		
2	Will you learn Groups without Teacher through Computer?	14	77.78%	4	22.22%		
3	Is there any problem in computer based learning at Groups?	4	22.22%	14	77.78%		
4	Are you face any problem in Google's Quiz online test?	1	5.56%	17	94.44%		
5	Any chance to face attention splash in e-learning for Groups?	3	16.67%	15	83.33%		
6	Do you know any Education based websites for Groups?	2	11.11%	16	88.89%		
7	Did you face any problem in Groups e-learning?	3	16.67%	15	83.33%		
8	Which instrument you like in e-learning? (like TV, Radio, Computer...)	No. of like comp-uter	% of like comp-uter	No. of like TV	% of like TV		
		18	100%	0	0%		
9	Which learning software is suitable for Groups in computer? (PPT, Video...)	No. of like PPT	% of like PPT	No. of like video	% of like video	No. of like both	% of like both
		4	22.22%	2	11.11%	12	6.67%
10	Which one of the test you like online test or offline test for Groups?	No. of like online	% of like online	No. of like offline	% of like offline		
		7	38.89%	5	27.78%	6	33.33%

Table 139: Analysis of Permutations, Rajiv Gandhi University, Arunachal Pradesh.

Sl. no.	*Questions*	*No. of Yes*	*% of Yes*	*No. of No*	*% of No*	*Colu-mn 7*	*Colu-mn 8*
1	Do you like to learn Permutations through e-learning?	18	100	0	0		
2	Will you learn Permutations without Teacher through Computer?	15	83.33%	3	16.67%		
3	Is there any problem in computer based learning at Permutations?	2	11.11%	16	88.89%		
4	Are you face any problem in Google's Quiz online test?	2	11.11%	16	88.89%		
5	Any chance to face attention splash in e-learning for Permutations?	3	16.67%	15	83.33%		
6	Do you know any Education based websites for Permutations?	3	16.67%	15	83.33%		
7	Did you face any problem in Permutations e-learning?	1	5.56%	17	94.44%		
8	Which instrument you like in e-learning? (like TV, Radio, Computer...)	No. of like comp-uter	% of like comp-uter	No. of like TV	% of like TV		
		17	94.44%	1	5.56%		
9	Which learning software is suitable for Permutations in computer? (PPT, Video...)	No. of like PPT	% of like PPT	No. of like video	% of like video	No. of like both	% of like both
		3	16.67%	8	44.44%	7	38.89%
10	Which one of the test you like online test or offline test for Permutations?	No. of like online	% of like online	No. of like offline	% of like offline		
		10	55.56%	5	27.78%	3	16.67%

Table 140: Analysis of Recurrence Relations, Rajiv Gandhi University, Arunachal Pradesh.

Sl. no.	*Questions*	*No. of Yes*	*% of Yes*	*No. of No*	*% of No*	*Colu-mn 7*	*Colu-mn 8*
1	Do you like to learn Recurrence relations through e-learning?	18	100%	0	0%		
2	Will you learn Recurrence relations without Teacher through Computer?	12	66.67%	6	33.33%		
3	Is there any problem in computer based learning at Recurrence relations?	1	5.56%	17	94.44%		
4	Are you face any problem in Google's Quiz online test?	3	16.67%	15	83.33%		
5	Any chance to face attention splash in e-learning for Recurrence relations?	3	16.67%	15	83.33%		
6	Do you know any Education based websites for Recurrence relations?	3	16.67%	15	83.33%		
7	Did you face any problem in Recurrence relations e-learning?	2	11.11%	16	88.89%		
8	Which instrument you like in e-learning? (like TV, Radio, Computer...)	No. of like computer	% of like computer	No. of like TV	% of like TV		
		17	94.44%	1	5.56%		
9	Which learning software is suitable for Recurrence relations in computer? (PPT, Video...)	No. of like PPT	% of like PPT	No. of like video	% of like video	No. of like both	% of like both
		4	22.22%	3	16.67%	11	33.33%
10	Which one of the test you like online test or offline test for Recurrence relations?	No. of like online	% of like online	No. of like offline	% of like offline		
		5	27.78%	8	44.44%	5	27.78%

Unique feature of this group is they were heterogeneous group in true sense. This is Central University therefore students were from different states like Arunachal Pradesh, Assam, Mizoram, Meghalaya, Tripura and Kerala. Moreover they were from different programmes as mentioned earlier. Their culture, race, religion and language were different. Most of them were from tribal background. But all understood common language Hindi and English. Although the group was heterogeneous it was amazing to see them learning together and their interaction with investigators was excellent. The atmosphere was very friendly and conducive. One of the students welcomed the investigators to visit his home in Tawang and told that it is as beautiful as Switzerland. In short it was most wonderful and unforgettable visit for the investigators.

Step 3

There were 18 students in the selected class in Rajiv Gandhi University. Their online feedback was analyzed and its results are as follows.

17.3.1. *Topic equivalence relation*

The following (Table 137) is the analysis of the questionnaire result for the topic Equivalence Relation.

There were six positively oriented questions and four negatively oriented questions more load (more than 50% of students) is on **YES** in positively oriented questions (questions 1, 2, 8, 9, 10) for positively oriented questions 6 more load in on **NO** side, more load (more than 50% of students) is on **NO** in negatively oriented questions (3, 4, 5, 7).

Analysis clearly shows that most of the students liked to learn the topic Equivalence Relation using e-learning material.

17.3.2. *Topic groups*

The following (Table 138) is the analysis of the questionnaire result for the topic Groups.

There were six positively oriented questions and four negatively oriented questions more load (more than 50% of students) is on **YES** in positively oriented questions (questions 1, 2, 8, 9, 10) for positively oriented questions 6 more load in on **NO** side, more load (more than 50% of students) is on **NO** in negatively oriented questions (3, 4, 5, 7).

Analysis clearly shows that most of the students liked to learn the topic Groups using e-learning material.

17.3.3. *Topic permutations*

The following (Table 139) is the analysis of the questionnaire result for the topic Permutations.

There were six positively oriented questions and four negatively oriented questions more load (more than 50% of students) is on **YES** in positively oriented questions (questions 1, 2, 8, 9, 10) for positively oriented questions 6 more load in on **NO** side, more load (more than 50% of students) is on **NO** in negatively oriented questions (3, 4, 5, 7).

Analysis clearly shows that most of the students liked to learn the topic Permutations using e-learning material.

17.3.4. *Topic recurrence relations*

The following (Table 140) is the analysis of the questionnaire result for the topic Recurrence Relations.

There were six positively oriented questions and four negatively oriented questions more load (more than 50% of students) is on **YES** in positively oriented questions (questions 1, 2, 8, 9, 10) for positively oriented questions 6 more load in on **NO** side, more load (more than 50% of students) is on **NO** in negatively oriented questions (3, 4, 5, 7).

Analysis clearly shows that most of the students liked to learn the topic Recurrence relations using e-learning material.

18. DISCUSSION

After analysing and interpreting the obtained data the investigators found that there is a significant gain in terms of achievement of the students who learnt discrete mathematics through e-learning material. Responses in the reaction scale was positive, students liked learning through e-learning material. Investigators observed that students enjoyed e-learning material and was stress free while they were learning. Contents were presented in variety of ways and care was taken to present the content in increasing order of difficulty. Visual images helped the students to understand content with ease. Variety of exercises helped students to be actively involved in the learning process. Online Videos and downloaded videos were presented to students for further understanding. Investigators recorded videos of discrete mathematics syllabus and the lecture was given by Mathematics Professor from Department of Mathematics Central University of Tamil Nadu. These recorded videos were shown to experimental group students. Students liked these videos and this helped them to consolidate their learning.

The investigators reviewed research related to CAI which was very much similar to e-learning to see the effectiveness. CAI helped students to enhance their learning. They also reviewed researches related to e-learning in subjects other than mathematics to see the effectiveness in those subjects and in most of the research e-learning helped students for better learning. Investigators narrowed the review to effectiveness of e-learning in mathematics. They found that e-learning was effective in mathematics. Further investigators narrowed the review to effectiveness of e-learning material in discrete mathematics and found that it was effective. The reviewed research is discussed below in detail.

In (Strother, 2002) study "An Assessment of the Effectiveness of e-learning in Corporate Training Programs" he found that e-learning was equally effective as classroom learning. (Andrew R. Brown, 2005) conducted a study on, "Elements of Effective E-learning Design" and found that students display high level of motivation because of immediate feedback in e-learning material. (Rita Birzina, 2009) found that use of ICT will be effective if it provides students experience and social recognition. (Umek *et al.,* 2015) found that there is positive co-relation between moodle based e-learning platform and students satisfaction. (Anshari *et al.,* 2015) found that e-learning in higher education can be helpful for flexible learning management. (Kenan, 2015) found that e-learning material can cater diverse learners. (Liu, Sheng, Hsueh & Sung-Lin, 2016) found that e-learning helped in transfer of learning and also will provide teachers in some assistance in teaching methods. (Kattoua, Al-Lozi & Alrowwad, 2016) concluded that a successful e-learning system should consider the personal, social, cultural, technological, organizational and environmental factors. (Zare *et al.,* 2016) study supports researchers and practitioners in effectively adopting MCDM techniques regarding E-learning evaluation. (Basogain, Olabe & Olabe, 2017) reviews the role of e-learning as an integral part of the transition from traditional education systems to modern systems. (Lubov *et al.,* 2017) shown that the use of 3D-models and the implementation of the visualizer mode helps to achieve better learning outcomes. (Sheshasaayee & Bee, 2017) paper reports, how the previous scholars tackled the earlier/recent trends in e-learning research. (Ren Lin & Kao, 2018) e-learning system can effectively facilitate users' self-awareness of mental efforts in online learning contexts. (Vallejo *et al.,* 2018) results showed significant differences in learning achievement and academic self-efficacy attributable to the effect of the scaffolding.

The following are the studies related to e-learning in mathematics. (Babor, Fassi, Hariri & Bideq, 2016) in their studies "An e-learning environment for algorithmic: Toward an active construction of skills". They concluded by presenting around the proposed problem a pedagogical scenario for the set of learning activities designed to be incorporated in an E-learning platform. (Borba, C. *et al.,* 2016) in this literature survey they focused on identifying recent advances in research on digital technology in the field of mathematics education. They concluded the report discussing some of the implications that these digital technologies may have for mathematics education research. (Bray & Tangney, 2017) found that lower levels of transformative usage of technology in mathematics education. (Miyazaki, 2018) studied about "Web-based Application to Aid the Learning of Mathematical Proofs Using Logical Diagrams and Its Potential". (Junaedi, Suyitno, Sugiharti & Eng, 2015) study discusses about problems in learning discrete mathematics.

Researches related to e-learning in discrete mathematics. (Escribano, Antonio, Carbajo & Rosa, 2008) opined that TICs (technologies for informatics and communication) provide a very positive aid to the learning tasks. Interactive tutorials allow access to a big amount of information in a multi-sequential way, by using specifically designed Java applets suitable for the learning of a specific knowledge. This facility helps students in acquiring in depth knowledge in learning discrete mathematics. (Kularbphettong, Putglan, Tachpetpaiboon, Tongsiri & Roonrakwit, 2015) their objective of this system is to advice student to learn discrete mathematics more convenient on mobile platform. (Kamèeva *et al.,* 2016) found that depending on the type of student, learning style, level of knowledge and dynamic behavior, a general conclusion is drawn that the students spend more time in non-adaptive *vs*. adaptive e-learning systems. (Hernández & González-Ramírez, 2017) their results of the research showed the impact of the e-textbook elaborated in the engagement towards the learning of Discrete Mathematics, (Ivanov & Saltan, 2017) in their study "Discrete mathematics course supported by CAS MATHEMATICA", the results of the survey provided evidence that their approach contributes to high outcomes and aligns with the course aims and objectives.

From the above discussion it is clear that e-learning material enhances the learning in most of the subjects especially mathematics and in particularly discrete mathematics. Investigators also found that e-learning material was effective in discrete mathematics in the selected four topics. This point is further strengthened by investigators observation of students' reaction during learning through e-learning. They were relaxed and were without stress during their e- learning phase. From this we can safely concluded that e-learning material is effective in learning selected topics in discrete mathematic.

Section 6

19. CONCLUSIONS

Discrete mathematics is very important topic in bachelor's degree in mathematics and all computer science related degrees. This study proves that e-learning motivates them to learn further and stress free learning. It also caters different type of learning style of students like slow learners, mediocre learners and fast learners they can study in their own respective speeds. E-learning helped students in developing self-esteem and self confidence that they can learn by themselves without the help of teachers. Students will be active learners whenever they want they can learn by themselves instead of passive learning in traditional classroom teaching. This e-learning material can also help in management of available space in universities and colleges because they can learn wherever they want and special classroom is not a must to learn this material.

20. TIME TAKEN FOR THE STUDY

It took one year to conduct the study.

21. IMPLICATIONS OF THE STUDY

An e-learning environment can supplement or complement a traditional face-to-face learning environment. It provides greater access to learning and provides greater diversification to learning. (Leong, 2006) E-learning is a sub-set of flexible teaching and learning that seeks to provide greater access to learning for all students. Flexible teaching seeks to provide educational programs in ways that meet the changing needs of learners. It enhances learning, empowers students, time saving, active learning, promotes self-learning and academic success. In future, perhaps, lecturers could introduce a variety of new e-learning subject deliveries and activities so as to engage students in the learning and hence make learning meaningful and enjoyable. Tertiary education should look beyond the traditional boundaries of classroom instruction by augmenting their current best practices with the new advances in learning and collaboration technologies to maximise results.

22. SUGGESTIONS FOR FURTHER SCOPE OF THE STUDY

The present study may bring new area to be studied by the future researchers. The content which was not covered by the present study can be taken up for further study. E-learning material can be developed in other topics in discrete mathematics which is not covered in the present study. All higher educational institutions have computer facilities and this facility can be used to learn subjects using e-learning material. E-learning material can be developed in other topics of mathematics offered at undergraduate level. E-learning can be developed with facility of on line discussion with the students.

23. ACKNOWLEDGEMENTS

This project is fully funded by ICSSR, Government of India. The completion of this study would have been not possible if not dependent on the steadfast support of ICSSR and Central University of Tamil Nadu (CUTN), Thiruvarur - 05. ICSSR and CUTN paid equal provision and contribution to the study for which we always feel profound gratitude in our heart.

We also owe our special thanks to the following members who helped us in collecting the data:

to **Dr. R. Sivakumar**, Associate Professor & Head of Department of Computer Science and the management of **A. Veeriya Vandayar Memorial Sri Pushpam College (Autonomous) Poondi** - 613 503;

to **Mrs. S.M. Valli,** Associate Professor & Head of Department of Computer Science and the Management of **Thiagarajar College of Arts and Science, Madurai** - 625009;

to **Dr. Mandira Sikdar**, Dean-Student Affairs, Associate Professor, Science, SLSE and the management **of Navrachana University**;

to **Mr. Ani Taggu**, Associate Professor and Head, **Mr. Firos A**, Assistant Professor, Department of Computer Science and Engineering, **Dr. Vivek Singh**, Assistant Professor, Department of Education from **Rajiv Gandhi Central University, Arunachal Pradesh**.

Last but not least we would like to thank the project staff without their support this project would not have been materialized.

24. REFERENCES

Abdi, H. and Williams, L.J. (2010). Tukey's Honestly Signiûcant Diûerence (HSD) Test. Retrieved 328, 2013, from *Encyclopedia of Research Design.*

Adhikari, R. (1992). Development of Computer Aided Instruction Material on cell and cell reproduction for class IX. *In*: Goel, D.R. (2000), Educational Media in India. New Delhi: Bharatiya Kala Prakashan, India.

Aguele, L. and O.O., A. (2010). Effectiveness of Selected Teaching Strategies in the Remediation of Process Errors Committed by Senior Secondary School Students in Mathematics. Retrieved from: *www.krepublishers.com/.../JSS-22-2-139-10-866-A*

Alberta Education, Alberta, Canada (2014). Mathematics Kindergarten to Grade 9. Retrieved February 16, 2015, from Alberta Education, Alberta, Canada: *http://education.alberta.ca/media/8775377/k_to_9_math_pos.pdf*

American Association for the Advancement of Science (1990). *Chapter 2: The Nature of Mathematics.* Retrieved 3 18, 2013, from Science for All Americans Online: *http://www.project2061.org/publications/sfaa/online/chap2.htm*

ANCOVA calculator used: *http://vassarstats.net/vsancova.html*

Andrew R. Brown, B.D. (2005,). The International Review of Research in Open and Distributed Learning. Retrieved February 16, 2015, from Athabasca University: *http://www.irrodl.org/index.php/irrodl/article/view/217/300*

Anshari Muhammad, Alas Yabit, Guan and Sei, L. (2015). Pervasive Knowledge, Social Networks, and Cloud Computing: E-learning 2.0. *Eurasia Journal of Mathematics,* 11(13). Retrieved from: *https://eric.ed.gov/?q=e-learning+in+discrete+mathematics&id=EJ1074058*

Anthony, G. and Walshaw, M. (2009). Characteristics of Effective Teaching of Mathematics: A View from the West. *Journal of Mathematics Education,* 2(2): 147–164. *http://educationforatoz.org/images/_9734_12_Glenda_Anthon on 21.10.2012.*

Babor, A., Fassi, H.F., Hariri, A. and Bideq, M. (2016). An e-learning environment for (111119). The Effect of Peer Tutoring on Achievement of Students in Discrete Structures. *Asia Pacific Journal of Multidisciplinary Research*, 3: 8–12. Retrieved from: *http://www.apjmr.com/wp-content/uploads/2016/08/APJMR-3.5.4.02.pdf*

Barot, H. (2005). To study the effectiveness of CAI in Sanskrit for std. VIII students. Vadodara: CASE. The Maharaja Sayajirao University of Baroda, India: An Unpublished M.Ed. Dissertation.

Basogain, X., Olabe, M. and Olabe, J. (2017). Transition to a Modern Education System through e-learning. Tutorial Research Refereed limited, Bangkok, Thailand, pp. 41–46. Retrieved from: *https://dl.acm.org/citation.cfm?id=3160924*

Bayturan, S. and Kesan, C. (2012). The Effect of Computer-Assisted Instruction on the Achievement and Attitudes towards Mathematics of Students in Mathematics Education.

Beaird (2007). The effects of computer-assisted language learning on English Language learners with and without disabilities in an elementary school setting. Ph.D. Thesis., University of Nevada, Las Vegas. *Dissertation Abstract International*, 68(4): 2947–A.

Beetham, H. (2005). E-learning research: Emerging issues? Retrieved from ALT-J, Research in Learning Technology: *http://files.eric.ed.gov/fulltext/EJ820152.pdf*

Bhatia, K. (1992). Identification and Remedy of Difficulties in Learning Fractions with Programmed Instructional Material. *In*: NCERT (*ed.*), *Fifth Survey of Research in Education*, New Delhi: NCERT.

Blanco, S. (2018). Active Learning in a Discrete Mathematics Class. ACM Digital Library, pp. 828–833. Retrieved from: *https://dl.acm.org/citation.cfm?id=3159450.3159604*

Borba, C., Askar, P., Engelbrecht, J., Gadanidis, G., Llinares, S. and Aguilar, M. (2016). Blended learning, e learning and mobile learning in mathematics education. *ZDM Mathematics Education,* 48: 589–610. Retrieved from: *https://bit.ly/2Kp8KPq*

Bray, A. and Tangney, B. (2017). Technology usage in mathematics education research a systematic review of recent trends. *Computers & Education*, 114(C): 255–273. Retrieved from: *https://www.sciencedirect.com/science/article/pii/S0360131517301677*

Brown, K.L. (2003). From teacher-centered to learner-centered curriculum: Improving learning in diverse classrooms. *Education*, 124(1), 49-54.

Campit, J., Cayabyab, J. and Galas, E. (2015). The Effect of Peer Tutoring on Achievement of Students in Discrete Structures. *Asia Pacific Journal of Multidisciplinary Research,* 3: 8–12. Retrieved from: *http://www.apjmr.com/wp-content/uploads/2016/08/APJMR-3.5.4.02.pdf*

Casanova, A. (2004). An Analysis of Computer Mediated Communication Technologies as Tools to Enhance Learning. *Dissertation Abstract International,* 65(12).

Chel, M. (1990). Diagnosis and remediation of under achievement in compulsary Maths of Madynamic examination in West Bengal. *In*: NCERT(1992), *Fifth Survey of Research in Education*. New Delhi: NCERT, India.

Crews, J.M. (2003). *Helping Poor Readers:A Case Study of Computer Assisted Instruction.* Retrieved 1 12, 2009 from: *http://www.jcrews @cmi. arizona. edu/*

Dalwadi, N. (2001). *Development of Computer Assisted Instruction in Science for the Students of Standard IX.* Vadodara: An unpublished M.Ed. dissertation, CASE, The Maharaja Sayajirao University of Baroda.

Dandapani, S. (2001). *A Textbook of Advanced Educational Psychology.* New Delhi : Anmol Publication.

Das, A. (1998). Exploring Effectiveness of Computer Assisted Learning Material on Rhymes in Different Modes. Vadodara: An unpublished Ph.D. Thesis, CASE, The M.S. University of Baroda.

Davies, J.L. (1982). Effects of different Modes of paring in Programmed Learning of Mathematics on the performance of Underachievers. *In:* Buch, M.B. (*ed.*), *Third Survey of Research in Education.* New Delhi: NCERT.

Deshmukh, S. and Deshmukh, V. (2017). Analysis of various Barriers to implementation of Web based Technology in education. *International Journal of Advanced Research in Computer Engineering & Technology (IJARCET),* 6(3): 214–217. Retrieved from: *http://ijarcet.org/wp-content/uploads/IJARCET-VOL-6-ISSUE-3-214-217.pdf*

Eck, D. (n.d.). *Chi-square Statistic.* Retrieved 3 19, 2013, from Mathbeans Project: *http://math.hws.edu/javamath/ryan/ChiSquare.html*

E-Learning Courses From the IITs and IISc. (n.d.). Retrieved December 29, 2014, from NPTEL: *http://nptel.ac.in/courses/103101111/2*

Escribano, C., Antonio, I., Carbajo, G. and Rosa, S. (2008). Interactive tools for Discrete Mathematics E-learning. *WSEAS Transactions,* 5(2): 96–102. Retrieved from: *https://bit.ly/2Knc3GZ*

Eteokleous, N.P. (2004). Computer Technology Integration in Cyprus Elementary Schools. *Dissertation Abstract International*, 65(12).

Fassi, H.F., Hariri, A. and Bideq, M. (2016). An e-learning environment for algorithmic: Toward an active construction of skills. *World Journal on Educational Technology,* 8(2): 82–90. Retrieved from: *https://files.eric.ed.gov/fulltext/EJ1141892.pdf*

Ford, G.S. (2007). Effect of computer-aided instruction versus traditional modes on student PT's learning musculoskeletal special tests. Ph.D. Thesis., State University of New york at Buffalo. *Dissertation Abstract International*, 68(12): 5020–A.

Galvis, A.T. (2007). Computer Assisted Instruction (CAI) as a Teaching Tool for Occupational Therapy Education: A guide to understand CAI design and effectiveness. Ph.D. Thesis, Texas Women's University. *Dissertation Abstract International.*

Garcia, R., Falkner, K. and Vivian, R. (2018). System Self-Regulated Learning strategies using E-learning tools for Computer Science. *Elsevier*, 123: 150–163. Retrieved from: *https://www.sciencedirect.com/science/article/pii/S0360131518301064*

Gay, L.R. (1973). *Educational Research.* New Jersey: Prentice Hall.

George, P. (1969). *"The goals of Mathematical education," in ComMuniCator.* Retrieved 1 15, 2013, from: *www.ncert.nic.in/html/pdf/schoolcurriculum/.../prelims.pdf*

Gilbert, D.W. (2006). Effectiveness of computer-assisted instruction blended with class-room teaching methods to acquire automotive psychomotor skills. Ph.D. Thesis., Southern Illinois University at Carbondale. *Dissertation Abstracts International*, 67(8): 2947–A.

Gujarat Government Gazette (2012). *Right of Children to Free and Compulsory Education Act, 2009.* Gandhinagar: Education Department Notification. Sachivalaya. Extra No. 54.

Gujarat Government Gazette (2012). *The Gujarat Government Gazette.* Retrieved 3 20, 2013, from Rules and Orders (Other than those published in part I, I-A and I-L) made by the Government of Gujarat under the Gujarat Acts.: *http://righttoeducation.in/sites/default/files/ccs_rte-rules_gujarat.pdf*

Helaiya, S. (2004). Developing and Implementation of CAI package for teaching statistics to B.Ed. Students. Vadodara: An unpublished M.Ed. dissertation. CASE. The Maharaja Sayajirao University of Baroda.

Hernández, A. and González-Ramírez, T. (2017). *Design and assessment of the impact of an e-textbook in the engagement towards the learning of Discrete Mathematics. In*: Doodero, J., Salz, M. and Rube, I. (*eds.*), *TEEM.* ACM New York. Retrieved from: *https://dl.acm.org/citation.cfm?doid=3144826.3145443*

Hsu, J.J. (1994). Computer Assisted Language Learning (CALL) the Effect of ELS Students' Use of Instructional Modification on Listing Comprehension. *Dissertation Abstracts International,* 55(4): 851.

Hung, Y. (2005). The Evaluation of a Technology-Aided Lecture Accompanied by a Set of Macroeconomics Computer Interactive Exercises in Macroeconomics for the Undergraduate Business Major in Taiwan. Idaho State University. *Dissertation Abstract International*, 66(7): 2473–A.

Inamdar, J.A. (1981). A study of the Effectiveness of the programmed learning strategy in the subject of Mathematics for Standard VII in relation to some Psychological Correlates. *In:* Buch, M.B. (*ed.*), *Third Survey of Research in Education.* New Delh: NCERT.INC.

Initiative, J.A. (2010). *Computer Assisted Instruction (CAI) software for maths.* Retrieved January Tuesday, 2013, from JobAccess an Australian

Government Initiative: *http://jobaccess.gov.au/Advice/ProductOrSolutionOne/Pages/CAI_software_for_maths.aspx*

Ivanov, A. and Saltan, A. (2017). Discrete mathematics course supported by CAS MATHEMATICA. *International Journal of Mathematical Education in Science and Technology,* 48(6): 953–963. Retrieved from: *https://bit.ly/2tIxOqa*

Jackson, K. and Dave, W. (2011). *International Journal of Curriculum and Instruction.* Retrieved 2 3, 2012, from The Effect of Computer-Assisted Instruction on Student's Attitudes and Achievement in Matrices and Transformations in Secondary Schools in Uasin Gishu District, Kenya: *http://www.muk.ac.ke/ijci/ ISSN 2077-4982 ©2011* Department of Curriculum, Instruction and Educational Media, Moi University.

Jain, S. and Burad, G. (1988). Low Results in Mathematics at Secondry Examination in Rajasthan. Independent Study. Udaipur: State Institute of Educational Research and training.

Jayasree, S. (1997). A diagnostic study of the difficulties of pupils of standard VIII in expanding using identities. Retrieved 9 25, 12, from Unpublished M.Ed. Thesis, University of Kerala, Thiruvananthapuram: *http://shodhganga.inflibnet.ac.in/bitstream/10603/5*

Jeyamani, P. (1991). Effectiveness of simulation modes of teaching through CAI. *In*: NCERT (1992), *Fifth Survey of Research in Education.* New Delhi: NCERT, India.

Junaedi, I., Suyitno, A., Sugiharti, E.and Eng, E. (2015). Disclosure Causes of Students Error in Resolving Discrete Mathematics Problems Based on NEA as A Means of Enhancing Creativity. *International Journal of Education,* 7: 31–42. Retrieved from: *http://www.macrothink.org/journal/index.php/ije/article/view/8462*

Kamèeva, Spasova E., Ilieva, G., Lidija Mitrevski, Pece and Sime (2016). *Adaptive Model for Discrete Math of* the E-leaerning System. 25: 405–411. Retrieved from: https:*//bit.ly/2lDlrYM*

Kapoor, J. (1997). Mathematics Education: A trend Report. *In:* NCERT(Ed.), *Fifth survey of Educational Research 1988-92* (Vol. 1). NCERT, New Delhi.

Karnati (2008). Computer aided instruction for out-of-school children in India: An impact study in Andhra Pradesh. Ph.D. Thesis., University of Pennsylvania. *Dissertation Abstract International,* 69(4): 1335–A.

Kasat, B. (1991). In-depth study of the causes of the large failures in mathematics at S.S.C. examination of Marathi medium high school students in Plaghar Tahsil. *In:* NCERT(1992), *Fifth Survey of Research in Education.* New Delhi: NCERT.

Kattoua, T., Al-Lozi, M. and Alrowwad, A. (2016). A Review of Literature on E-learning Systems in Higher Education. *International Journal of*

Business Management and Economic Research, 7(5): 754–762. Retrieved from: *file:///E:/Downloads/AReviewofLiteratureonE-Learning%20(1).pdf*

Kenan, T. (2015). Improving the Effectiveness of E-learning Implementation in the School of Engineering at Tripoli University. Tripoli University. Huddersfield: The University of Huddersfield. Retrieved from: *http://eprints.hud.ac.uk/id/eprint/28702/1/Final%20thesis%20-%20KENAN.pdf*

Khirwadkar, A. (1998). Development of Computer software for learning chemistry at standard XI. Vadodara: CASE, The M.S.University of Baroda, India: An unpublished Ph.D, Thesis.

Kinney, D. and Robertson, D.E. (2003). Technology makes possible new models for delivering developmental mathematics instruction. *Mathematics and Computer Education,* 37(3): 315–328.

Kiravu Cheddi, Yanev, Kamen M., Tunde, Moses O. and Ansel. (2016). E-learning Hands-On: Blending Interactive E-learning with Practical Engineering Laboratory. *International Journal of Information and Learning Technology,* 33(22). Retrieved from: *https://bit.ly/2KciQUJ*

Kiravu, C. (2016). E-learning hands-on: Blending interactive E-learning with practical engineering laboratory. *International Journal of Information and Learning Technology,* 33(5). Retrieved from: *https://www.emeraldinsight.com/doi/abs/10.1108/IJILT-07-2016-0023*

Kirkman, T. (1996). ANOVA Analysis of Variance between groups. Retrieved 10 9, 2012, from Statistics in Use: *http://www.physics.csbsju.edu/stats*

Koul, L. (2008). *Methodology of educational Research.* Shimla: Vikas Publishing House Pvt. Ltd.

Kularbphettong, K., Putglan, R., Tachpetpaiboon, N., Tongsiri, C. and Roonrakwit, P. (2015). Developing of m-learning for Discrete Mathematics based on Android Platform. *Procedia - Social and Behavioral Sciences.* 197: 793–796. Bangkok, Thailand: ELSEVIER. Retrieved from: *https://core.ac.uk/download/pdf/81953710.pdf*

Kulkarani, P.V. and Yadav, M.S. (1966). A comparative Study of teaching by Different Methods of Programming of Different Levels of Pupils. *In*: Buch, M. *First Survey of Research in Education* (1970 ed.). Vadodara: Society for Educational Research and Development.

Leong, H.T. (2006). *2nd International CDIO Conference.* Retrieved from Implications of E-learning on Learning and Teaching in Higher Education : *http://www.cdio.org/files/document/file/HoTeckJune.pdf*

Liu, Sheng K., Hsueh and Sung Lin (2016). Effects of Digital Teaching on the Thinking Styles and the Transfer of Learning of the Students in Department of Interior Design. *EURASIA Journal of Mathematics,*

Science & Technology Education, 12: 1697–1706. Retrieved from: *https://eric.ed.gov/?q=e-learning+in+discrete+mathematics&id=EJ1097650*

Lubov, S., Lisitsyna, Evgenii A., Efimchik, Svetlana A. and Izgareva (2017). RLCP-Compatible Virtual Laboratories with 3D-Models and Demonstration Mode: Development and Application in E-learning. Smart Innovation, Systems and Technologies, pp. 75–81. Russia: ITMO University. Saint Petersburg, Russia. Retrieved from: *https://link.springer.com/chapter/10.1007/978-3-319-59451-4_8*

McGraw-Hill Dictionary, O. S. (2003). Computer Assisted Instruction. Retrieved 3 18, 2013, from The free Dictionary: *http://encyclopedia2.thefreedictionary.com/computer-assisted+instruction*

McLaughlin, D. (2004). Towards a new Paradigm for Teaching and Learning: A Case Study of the Process of Integrating Instructional Design and Technology at Florida Community College at Jacksonville. *Dissertation Abstract International*, 65(10): 3667.

Miyazaki, Y. (2018). Web-based Application to Aid the Learning of Mathematical Proofs Using Logical Diagrams and Its Potential. *Proceedings of Society for Information Technology & Teacher Education International Conference*, pp. 965–968. Association for the Advancement of Computing in Education. Retrieved from: *https://www.learntechlib.org/p/171805/*

Moila, M. (2006). The Use of Educational Technology in Mathematics Teaching and Learning: An Investigation of South African Rural Secondary School. on 21.10.2012: *http://upetd.up.ac.za/thesis/available/etd-04252008-143722/unrestricted/dissertation.pdf on 21.10.2012.*

Nagar, N. (1988). Effectiveness of Computers in Teaching Mathematics in Schools. *In:* NCERT (Ed.), *Fifth Survey of Research in Education* (1992 ed.). New Delhi: NCERT.

NCERT (1971). *Education and National Development. Report of the Education Commision 1964-66.* New Delhi: NCERT.

NCERT (2000). *Report of the Education Commission 1964-66.* New Delhi: NCERT.

NCERT (2005). *Position Papers of National Focus Groups on Curricular Areas.* New Delhi: NCERT Volume 1.

Nimtrakul. (1999). Effectiveness of Computer Assisted Instruction on Atomic Structure in Chemistry. Retrieved 11 20, 12, from: *http://www.thaiedresearch.org/thaied/sites/all/modules/thaied/thaied-results.tpl.php?-table=thaied_results&-action=browse&-cursor=1464&-skip=1440&-limit=30&-mode=list&-sort=id+asc&-recordid=thaied_results%3Fid%3D1468*

NPE (1992). *National Policy on Education*. Retrieved January 15, 2013, from NCERT: *http://www.ncert.nic.in/oth_anoun/npe86.pdf*

Pandey, I.D. (1980). Use of Programmed Instruction on Teaching Mathematics at Primary Level. *In:* Buch, M.B. (*ed.*), *Third Survey of Research in Education*. New Delhi: NCERT.

Pardeshi, R. (2005). *A study of the relative effectiveness of CAI and CAIPI in earning Trigonometry by English medium students of Standard IX of Baroda City*. Retrieved 10 26, 2012, from http://www.educationinindia.net/download/Research_Abstracts.pdf

Parikh, P.D. (2006). *Developing and Implementing Computer Assisted Learning Material for 11th std Commerce Students on Subject Introduction to Book-keeping and Accountancy Prescribed by GSEB*. Vadodara: An unpublished M.Ed. dissertation. Vadodara: CASE. The Maharaja Sayajirao University.

Patel, C.B. (1975). To Develop Auto instructional Programmes in some Units of Gemmetry for Class IX and to Find out their Effectiveness in relation to Different Variables. *In:* Buch, M.B. (*ed.*), *Third Survey of Research in Education*. New Delhi: NCERT.*Second Survey of Research in Education* (1970 ed.). Vadodara: Society of Education Research and Development.

Patel, A. (1977). Development and Tryout of Auto instructional Programmes in some Units of Gemmetry for Class VIII and to Study its Effectiveness in the Context of Different Variables. *In:* Buch, M.B. (*ed.*), *Second Survey of Research in Education* (1970 ed.). Vadodara: Society for Educational Research and Development.

Patel, J.A. (2009). *Development and Implementation of CAI to Teach English Grammar to Standard VIII Student in Different Modes*. Retrieved 10 27, 2012, from shodhganga.inflibnet.ac.in/bitstream/10603/.../09_chapter%202.pdf

Patel, K. (2008). *Computer Assisted Instruction in Physics for the Students of Standard XI*. Retrieved 10 27, 2012, from shodhganga.inflibnet.ac.in/bitstream/10603/.../09_chapter%202.pdf

Patel, M. (2006). *Factors Influsing Higher Secondary Students for Selection of Commerce and Rejection of Science Stream in Vadodara City*. Vadodara: An Unpublished M.Ed. Dissertation, CASE, The M.S.University of Baroda.

Patel, R. (2001). *A study of learning through Computer Assisted Learning Material in relation to selected production variables and contiguity*. Vadodara: An Unpublished M.Ed Dissertation: CASE,The Maharaja Sayajirao University of Baroda.

Pathak, R. and Chaudhary, J.I. (2012). *Educational Technology*. (Das, Y. (*ed.*)) New Delhi: Dorling Kindersley(India) Pvt. Ltd.

Pilli, O. (2008). Retrieved 4 11, 2009, from The Effects of Computer-Assisted Instruction on the Achievement, Attitudes and Retention of Fourth Grade Mathematics Course: *https://etd.lib.metu.edu.tr/upload/3/12609336/index.pdf*

Platz, M., Krieger, M., Niehaus, E. and Winter, K. (2018). Suggestion of an E-proof Environment in Mathematics Education. *ICME-13 Monographs*, pp. 107–120. Retrieved from: *https://link.springer.com/chapter/10.1007/978-3-319-73748-5-8*

Pramila, G. (2014). Unpublished Ph.D. Thesis. Vadodara, India: The M S University of Baroda, Babor, A., Bayturan, S. and Kesan, C. (2012). The Effect of Computer-Assisted Instruction on the Achievement and Attitudes Towards Mathematics of Students in Mathematics Education.

Project (2000). *Computer Assisted Language Learning.* Retrieved 7 21, 2012, from *Chapter 2: Computer Assisted Instruction and Learning*: *http://www.computing.dcu.ie/~mward/mthesis/chapter2.pdf*

Ramani, P. and Patadia, H. (2012). Comparative Analysis of Reaction of Students on Final Version of Computer Assisted Instruction for Teaching Arithmetic with Different Modes. *IOSR Journal of Humanities and Social Science (JHSS)*, 5(3): 52–91.

Ramani, P. and Patadia, H. (2012). Development and tryout of the Programmed Learning Material in Mathematics for class XI students studying in schools affiliated to Gujarat Secondary and Higher Secondary Education Board (GSHSEB). *Journal of Teacher Education in Developing Nations,* 1(5): 15.

Ramani, P. and Patadia, H. (2012). The effectiveness of Computer Assisted Instruction in Teaching Arithmetics. *International Journal of Scientific and Research Publications*, 2(11): 52–91.

Ramani, P. and Patadia, H. (2012,). Computer Assisted Instruction in Teaching of Mathematics. *IOSR Journal of Humanities and Social Science (JHSS),* 2(I): 39–42.

Ramani, P. and Patadia, H. (2013). Reaction of Students on Developed Computer Assisted Instruction for Teaching Arithmetic. *Scientific and Academic Publication*, 3(1).

Rao, T.G. (1983). A Comparative Study of Programmed Learning and Conventional Learning Methods in One Instruction of Mathematics:A Psychological Approach. *In:* Buch, M.B. (*ed.*), *Fourth Survey of Research In Education.* New Delhi: NCERT.

Rathwa, M. (2007). *Development and Implementation of Multimedia Package for teaching Gujarati subject.* Vadodara: An unpublished M.Ed. dissertation. CASE. The Maharaja Sayajirao University of Baroda.

Ravindra, G. (2006). *Researches on Curriculum and Teaching Mathematics.* New Delhi: NCERT.

Ren Lin, F. and Kao, C.M. (2018). Mental effort detection using EEG data in E-learning contexts. *Computers & Education,* 122: 63–79. Retrieved from: *https://www.sciencedirect.com/science/article/pii/S0360131 51830 0794*

Report of the Education Commission (1964-1966). Education and National Development 1964-1966. New Delhi: Ministry of Education: Government of India.

Retrieved 9 15, 2012, from Unpublished M.Ed Thesis. University of Kerala: *http://shodhganga.inflibnet.ac.in/bitstream/10603/564/9/09_chapter3.pdf*

Retrieved from: *https://www.sciencedirect.com/science/article/pii/S1568494616301685*

Rita Birzina, D.K. (2009). *Effectiveness of interactive e-learning organization and quality*. Retrieved February 16, 2015, from Education Line: *http://www.leeds.ac.uk/educol/documents/186734.pdf*

Robkob (1999). *Achievement and Retention in Science of Prathom Suska 5 Students in Science Studying through CAI*. Retrieved 6 6, 2012, from http://www.thaiedresearch.org/thaied/sites/all/modules/thaied/thaied-results.tpl.php?-table=thaied_results&-action=browse&-cursor=1585&-skip=1560&-limit=30&-mode=list&-sort=id+asc&-recordid=thaied_results%3Fid%3D1590

Rohayani, H., Kurniabudi and Sharipuddin (2015). Readiness Factors to measuring E-learning Readiness in Higher Education. *Science Direct,* 59: 230–234. Indonesia: Elsevier. Retrieved from: *https://pdfs.semanticscholar.org/b73d/e5e8214a61b4a6cddcc19718d18e86570277.pdf*

Rosales, J.S. (2005). The effect of computer-assisted instruction on mathematics achievement of ninth-grade high school students in the lowere Rio Grande Valley. Ed. University of Houston. *Dissertation Abstract International*, 66(7): 2482–A.

Rose Antony Stela, V. (1992). Effectiveness of Computer Assisted Instruction with special reference to underachievers. Ph.D. Education Bharathidasan University. New Delhi: NCERT, India: *Fifth Survey of Research in Education*, NCERT.

Ruttanathummatee, S. (2004). *Effectiveness of Computer Assisted Instruction for Primary School Students: An Experimental Study*. Retrieved 10 2012, 2012, from: *http://www.educationinindia.net/download/Research_Abstracts.pdf*

Salsbury, D.E. (2002). A study on comparing teacher-directed and Computer-Assisted Instruction of Elementary Geographic Place Vocabulary. Ph.D. Thesis, Kansas State University. *Dissertation Abstracts International*, 63(7): 2449–A.

Sarva Shiksha Abhiyan (2012). *Statistics and Survey*. Retrieved 3 20, 2013, from Sarva Shiksha Abhiyan: *http://www.ssa.tn.nic.in/Statistics.htm*

Sashidharan, T. (1992). *Learning intellectual skills as an educational outcome in relation to students entry characteristics and quality of instruction.* Vadodara: An unpublished Ph.D. thesis, CASE, M.S. University of Baroda.

Seshadri, M. (1980). An Experiment in the Use of Programmed Instruction in Secondary Schools. *In:* Buch, M.B. (*ed.*), *Third Survey of Research in Education.* New Delhi: NCERT.

Shah, J.C. (1981). To Develop and Try Programmed Material in Mathematics for students of Class V in Gujarat State. *In:* Buch, M.B. (*ed.*), *Third Survey of Research in Education.* New Delhi: NCERT.

Shah, M.S. (1969). To develop Auto-instructional Programmes in Algebra for Standard VIII. *In:* Buch, M.B. *First Survey of Research in Education* (1970 ed.). Vadodara: Society for Educational Research and Development.

Sharma, D. (2003). *A study of the effectiveness of Computer Assisted Learning (CAL) in chemistry for the students of standard XI.* Vadodara: An unpublished M.Ed. dissertation: CASE. The Maharaja Sayajirao University of Baroda.

Sharma, M.M. (1966). A Comparative Study of Outcomes of Teaching of Algebra by Conventional Classroom method and Method of Programmed Instruction. *In*: Buch, M.B. *First Survey of Research in Education* (1970 ed.). Vadodara: Society for Educational Research.

Sharma, R. (2005). *Encyclopedia of teaching methodology* (Vol. 1). New Delhi: Shubhi Publication.

Sheshasaayee, A. and Bee, N. (2017). E-learning: Mode to Improve the Quality of Educational System. *Smart Computing and Informatics*, *78*: 559–566. Chennai, India. Retrieved from: *https://link.springer.com/chapter/10.1007/978-981-10-5547-8_58*

Sidhu, K. (1995). *The teaching of Mathematics.* New Delhi: Sterling Publishers Private Limited.

Singh, A.K. (1982). *Tests Measurements and Research Methods in Behavioural Sciences.* New Delhi: Tata McGraw-Hill Publishing Company Limited.

Singh, R D. (1992). To Study the Effectiveness of Teaching Mathematics through Computer Assisted Instruction and Conventional Method of Cognitive and Non Cognitive Variables. *In*: NCERT (Ed.), *Fifth Survey of Research in Eduction.* New Delhi: NCERT.

Singh, R.D., Ahluwalia, S.P. and Verma, S.K. (1991). *Effectiveness of Computer Assisted Instruction (CAI) and Conventional method of instruction.* Retrieved 10 10, 2012, from: *shodhganga.inflibnet.ac.in/*

*bitstream/10603/.../09_chapter%202.pdf*socscistatistics. (n.d.). Retrieved May 8, 2018, from: *http://www.socscistatistics.com/tests/goodnessoffit/Default2.aspx*

Spradlin, K.D. (2010). *The Effectiveness of Computer- Assisted Instruction in Developmental Mathematics*. Retrieved 10 12, 12, from Faculty Publications and Presentations: *http://digitalcommons.liberty.edu/educ_fac_pubs/195*

Stella, V. R. (1992). *Effectiveness of Computer Assisted Instruction with Special reference to underachievers. In NCERT(1992).* New Delhi: NCERT.

Strother, J.B. (2002). An Assessment of the Effectiveness of E-learning in Corporate Training Program. Retrieved February 16, 2015, from Athabasca University: *http://www.irrodl.org/index.php/irrodl/article/viewArticle/83/160Tauber*

Suthar, K.S. (1981). A Study of Performance on Programmed Learning Material in Relation to Some Psychological Characteristic. *In*: Buch, M.B. (*ed.*), *Third Survey of Research in Education.* New Delhi: NCERT.

Suwanma (1991). Retrieved from Construction of Computer Assisted Instruction in science on topic "Earth and Changing" for Mathoyom Suska 2: *http://www.chinagmai.ac.th*

Tare, J. (2001). A Study of the Effectiveness of Branching Variety of Programmed Instructional Material as Diagnostic and Remedial Tool in Chemistry for Secondary Classes in Jabalpur Division. Retrieved 10 23, 2012, from: *http://www.educationinindia.net/download/Research_Abstracts.pdf*

Thakkar, S.I. (2006). To Develop and Implement CAI for Organization of Commerce and Management' Subject in Standard XI as Prescribed by GSEB. Vadodara: An unpublished M.Ed. dissertation.CASE. The Maharaja Sayajirao University of Baroda.

Thatte, C.H. (1998). An Experimental Study of the Relative Effectiveness of Programmed Learning and Learning through Audio Visual Aids with reference to certain selected topics from the syllabus of Science for Std. V to VII in Greater Bombay. Retrieved 9 25, 2012, from Ph.D. Thesis, University of Mumbai: *http://www.educationinindia.net/download/Research_Abstracts.pdf*

The free Dictionary (2008). Computer Assisted Instruction. Retrieved 3 18, 2013, from: *http://encyclopedia2.thefreedictionary.com/computer-assisted+instruction*

The National Research Council (2001). Adding it up: Helping children learn mathematics. Centre of education. Washington, DC: National Academy Press.

Tools for science. (n.d.). Retrieved 5 2, 2011, from: *http://www.physics.csbsju.edu/cgi-bin/stats/chi-square_form.sh?k=5*

Trivedi, I.U. (1980). Use of Branching Variety of Programmed Learning Materials as Diagnostic and Remedial Tools. *In*: Buch, M.B. (*ed.*), *Third Survey of Research in Education*. New Delhi: NCERT.

Tuckman, B.W. (1972). *Conducting Educational Research*. New York: Harcourt Brace Jovanovich,

Umek Lan, Aristovnik Aleksander, Tomaževic Keržic and Damijana (2015). Analysis of Selected Aspects of Students' Performance and Satisfaction in a Moodle-Based E-learning System Environment. *EURASIA Journal of Mathematics,* 11: 1495–1505. Retrieved from: *https://eric.ed.gov/?q=e-learning+in+discrete+mathematics&id=EJ1078244*

Vaisopha (1999). Construction of Computer Assisted Instruction in the Mathematics on topic 'Adding fraction' for Prahom Suska 5 students. Retrieved 7 21, 2012, from: *http://www.chinagmai.ac.th/abstract199921.html/ York : Bantam Books*

Vallejo, V., Nilson, Vargas L., Omar, Rodríguez S. and Luis. (2018). Effect of Motivational Scaffolding on E-learning Environments: Self-Efficacy, Learning Achievement and Cognitive Style. *Journal of Educators Online,* 15(14). Retrieved from: *https://eric.ed.gov/?q=e-learning+in+discrete+mathematics&id=EJ1168944*

Vansia (2011). Effectiveness of Computer with Peer Interaction for Math's learning in urban area. Retrieved 7 13, 2012, from *International Referred Research Journal*, ISSN- 097-3486:Rajbil 2001/30097, Vol-II Issue 24: *http://www.ssmrae.com/admin/images/f76c407be2bb7e5c6eef3071f8c9abd5.pdf*

Vansia (2011). Effectiveness of Computer with Peer Interaction for Math's learning in urban area. Retrieved 7 13, 2012, from: *International Referred Research Journal*, II (24). ISSN- 097-3486:Rajbil 2001/30097 *http://www.ssmrae.com/admin/images/f76c407be2bb7e5c6eef3071f8c9abd5.pdf*

Vasanthi, A. and Hema, S. (2003). Effectiveness of teaching Chemistry for 1 year B.E. students through Computer Assisted Instruction. Retrieved 11 12, 2012, from: *shodhganga.inflibnet.ac.in/bitstream/10603/.../09_chapter%202.pdf*

Vasudevan, A. (2003). *A Diagnostic study of the difficulties of pupils in Standard VIII in the computation of negative, numbers.*

Wagh, S. (1991). Development af Multi-media Instructional System for Remedial Measures in Fractional Numbers. Retrieved 10 25, 2012, from Shivaji Univesity India, Ph.D. Thesis in Education: *books.google.co.in/books?isbn=8183562094*

Wheeler, D. (1982). *Mathematization Matters*. Retrieved 1 15, 2013, from: *www.ncert.nic.in/html/pdf/schoolcurriculum/.../prelims.pdf*

Wikimedia Foundation, I. (2012). *wikipedia*. Retrieved October 18, 2012, from en.wikipedia.org: *http://en.wikipedia.org/wiki/Mathematical_economics*

Wikipedia (2013). *Analysis of variance*. Retrieved February 1, 2013, from Wikipedia The Free *Encyclopedia: http://en.wikipedia.org/wiki/Analysis_of_variance*

Wikipedia (n.d.). *E-learning*. Retrieved December 29, 2014, from Wikipedia: *http://en.wikipedia.org/wiki/E-learning*

Wikipedia (n.d.). *Piaget's theory of cognitive development*. Retrieved 7 13, 2012, from: *http://en.wikipedia.org/wiki/Piaget%27s_theory_of_cognitive_development*

Wikipedia Foundation, I. (2012). *wikipedia*. Retrieved October 18, 2012, from en.wikipedia.org: *http://en.wikipedia.org/wiki/Mathematical_economics*

Yadav, S. (2000). A study of the effectiveness of the Computer Software for students of standard I. Vadodara: An unpublisheAn Unpublished M.Ed. dissertation. Vadodara: CASE. The Maharaja Sayajirao University of Baroda.

Yasoda, R. (2009). *Problems in teaching and learning mathematics* (1st ed.). New Delhi: Discovery Publishing House Pvt. Ltd.

Zare, M., Pahl, C., Rahnama, H., Nilashi, M., Mardani, A., Ibrahim, O. and Ahmadi, H. (2016). Multi-criteria decision making approach in E-learning: A systematic review and classification. *Applied Soft Computing*, 45, 108–128.

Zyoud, M.M. (1999). Development of Computer Assisted English Language Teaching of VIII Standard Students. The Maharaja Sayajirao University of Baroda, India, CASE. Vadodara: An unpublished Ph.D. Thesis.

25. APPENDICES

25.1. Appendix 1: Photographs Taken During Experimentation

Photo 1: Data Collection in Central University of Tamil Nadu

Photo 2: Data Collection in Central University of Tamil Nadu

Photo 3: Data Collection in Central University of Tamil Nadu

Photo 4: Data Collection in Central University of Tamil Nadu

Photo 5: Data Collection in Central University of Tamil Nadu

Photo 6: Data Collection in Rajiv Gandhi Central University, Arunachal Pradesh

Photo 7: Data Collection in Rajiv Gandhi Central University, Arunachal Pradesh

Photo 8: Data Collection in Rajiv Gandhi Central University, Arunachal Pradesh

Photo 1

Photo 2

Photo 3

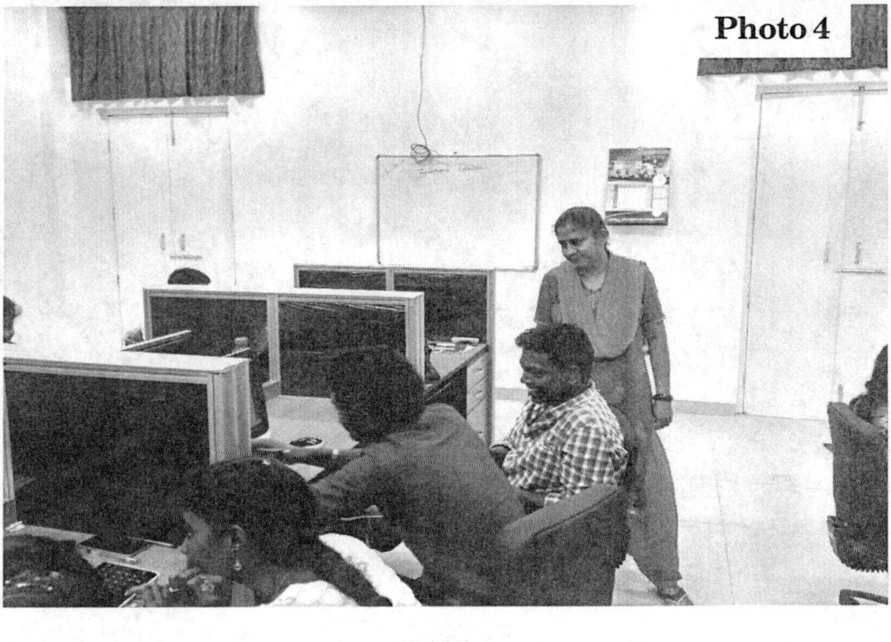

Photo 4

Photo 5

Photo 6

Photo 7

Photo 8

25.2. Appendix 2: Letter Given by Universities/Colleges for Conducting Experimentation

Thiagarajar College

(Re-Accredited with 'A' Grade by NAAC)
An Autonomous Institution affiliated to Madurai Kamaraj University
Post Box No. 107, 139-140, Kamarajar Salai, Teppakulam
Madurai - 625 009. TAMILNADU

Ph : 0452 - 2311875 Fax : 0452 - 2312375 Web : www.tcarts.in Email : principaltcarts@gmail.com

Dr. D. Pandiaraja
Principal

Date: 20.02.18

Certificate of Appreciation

This is to certify that **Dr. R. Pramila,** Assistant Professor, Department of Education, and **Dr. P. Thiyagarajan,** Assistant Professor & Head in-charge, Department of Computer Science, Central University of Tamil Nadu, Thiruvarur, has implemented the e-learning material, titled **Development and Implementation of e-learning in Discrete Mathematics** for the project sponsored by ICSSR. They also collected online feedback from first year students of B.Sc. Computer Science..

Dr. R. Pramila and Dr. P. Thiyagarajan also delivered keynote address on **"e-learning-materials"** in the programme organized by Department of Computer Science, Thiagarajar College, Madurai on 20th February 2018.

Principal

NAVRACHANA UNIVERSITY Reinventing Higher Education

NAVRACHANA UNIVERSITY

To Whom-so-ever Concerned

This is to certify that Dr. Pramila Ramani, Assistant Professor in Education, Central University of Tamil Nadu, visited the School of Liberal Studies and Education, Navrachana University, Vadodara, for implementation of the e-learning material she has designed for the project titled '**Development and Implementation of E-learning in Discrete Mathematics' (F.No:02/416/2016-17/RP)** sponsored by ICSSR for B.Sc. Sixth Semester students from 8th January to 12th January 2018.

Navrachana University Vadodara

Dr. Mandira Sikdar
Coordinator - B.Sc. Program,
SLSE, NUV.
10/1/2018.

Vasna-Bhayli Road, Vadodara-391 410, Gujarat, India. Tel.: 91 265 3020100 Tel/Fax : 91 265 3020110
Email : nuv@nuv.ac.in Website : www.nuv.ac.in

राजीव गांधी विश्वविद्यालय

(केन्द्रीय विश्वविद्यालय)

रोनो हिल्स, पी. ओ. दोईमूख -७९१११२

(अरुणाचल प्रदेश)

RAJIV GANDHI UNIVERSITY

(A Central University)

Rono Hills, P.O. Doimukh - 791 112

(Arunachal Pradesh)

Dated: 21.08.2018

TO WHOSOEVER IT MAY CONCERN

This is to certify that **Dr. R. Pramila** , Department of Education , Central University of Tamil Nadu and **Dr. P. Thiyagarajan** , Department Of Computer Science , Central University of Tamil Nadu delivered a lecture on " Basics of E-learning " and distributed the e-learning materials (Permutations, Groups, Recurrence Relation and Equivalence Relation) at Department of Computer Science and Engineering, Rajiv Gandhi University , Arunachal Pradesh on 20th August 2018.

We believe that the speech provided an excellent opportunity for the attendees as it provided both the theoretical background and principles behind "E-learning".

We believe that the technology you explained could provide a great level of support to the participants on their research work.

The feedback on the above modules received from the students in the form of questionnaire.

Again, thank you so much for your enthusiastic lecture and participation in our institution. We have no doubt that it would not have been the success that it was without your presence.

Best Regards,

21/8/2018 (FIROS.A)

Coordinator

21/08/2018

HOD.

Dept of CSE, RGU

Tel: +91-360-2277252 (Off), Fax: +91-360-2277317

25.3. Appendix 3: Reaction Scale Developed by the Investigators

Instructions for Reaction Scale

Dear students,

Thanks for your cooperation in participation of the experiment conducted in your university/college for my research project entitled "Development and Implementation of E-learning in Discrete Mathematics".

Please give your feedback regarding e-learning used during the experiment. Read the instructions before filling the reaction scale.

Read the sentence fully and understand what the sentence says. Put tick mark in one of the boxes

SA	A	ND	DA	SD

Given like these against each sentence (whichever you find appropriate about that sentence)

While doing this keep in mind the meaning for the following abbreviations.

SA - Strongly agree

A - Agree

ND - Not decided

DA - Disagree

SD - Strongly disagree

Investigators

(Dr. Pramila Ramani) (Dr. P.Thiyagarajan)

Central University of Tamil Nadu
Neelakudi Campus
Kangalancherry Post
Thiruvarur
Tamil Nadu
Pin code: 610005

1. Reaction Scale for Recurrence Relations

Sl. no.	*Questions*	*SA*	*A*	*ND*	*D*	*SD*
1	I like to learn Recurrence relations through e-learning.					
2	I can learn Recurrence relations without Teacher and only through e-learning material.					
3	I face problem in computer based learning at Recurrence relations.					
4	I face problem in answering Google's Quiz online test in Recurrence relations.					
5	I face attention splash in e-learning for Recurrence relations.					
6	I know Education based websites for Recurrence relations.					
7	I face problem in learning Recurrence Relations through e-learning material.					
8	I get boredom in e-learning on learning Recurrence Relations.					
9	I need teacher's help while learning Recurrence Relations through e-learning.					
10	I enjoyed e-learning as I can watch video in the topic Recurrence Relations					

SA: Strongly Agree, A: Agree, ND: Not Decided, D: Disagree, SD: Strongly Disagree

2. Reaction Scale for Permutations

Sl. no.	*Questions*	*SA*	*A*	*ND*	*D*	*SD*
1	I like to learn Permutations through e-learning.					
2	I can learn Permutations without Teacher and only through e-learning material.					
3	I face problem in computer based learning at Permutations.					
4	I face problem in answering Google's Quiz online test in Permutations.					
5	I face attention splash in e-learning for Permutations.					
6	I know Education based websites for Permutations.					
7	I face problem in learning Permutations through e-learning material.					
8	I get boredom in e-learning on learning Permutations.					
9	I need teacher's help while learning Permutations through e-learning.					
10	I enjoyed e-learning as I can watch video in the topic Permutations					

SA: Strongly Agree, A: Agree, ND: Not Decided, D: Disagree, SD: Strongly Disagree

3. Reaction Scale for Groups

Sl. no.	*Questions*	*SA*	*A*	*ND*	*D*	*SD*
1	I like to learn Groups through e-learning.					
2	I can learn Groups without Teacher and only through e-learning material.					
3	I face problem in computer based learning at Groups.					
4	I face problem in answering Google's Quiz online test in Groups.					
5	I face attention splash in e-learning for Groups.					
6	I know Education based websites for Groups.					
7	I face problem in learning Groups through e-learning material.					
8	I get boredom in e-learning on learning Groups.					
9	I need teacher's help while learning Groups through e-learning.					
10	I enjoyed e-learning as I can watch video in the topic Groups.					

SA: Strongly Agree, A: Agree, ND: Not Decided, D: Disagree, SD: Strongly Disagree

4. Reaction Scale for Equivalance Relations

Sl. no.	*Questions*	*SA*	*A*	*ND*	*D*	*SD*
1	I like to learn Equivalence Relations through e-learning.					
2	I can learn Equivalence Relations without Teacher and only through e-learning material.					
3	I face problem in computer based learning at Equivalence Relations.					
4	I face problem in answering Google's Quiz online test in Equivalence Relations.					
5	I face attention splash in e-learning for Equivalence Relations.					
6	I know Education based websites for Equivalence Relations.					
7	I face problem in learning Equivalence Relations through e-learning material.					
8	I get boredom in e-learning on learning Equivalence Relations.					
9	I need teacher's help while learning Equivalence Relations through e-learning.					
10	I enjoyed e-learning as I can watch video in the topic Equivalence Relations.					

SA: Strongly Agree, A: Agree, ND: Not Decided, D: Disagree, SD: Strongly Disagree

Subject Index